Copyright © 2020 by Zachary Schertz.

ISBN-000-0-0000-0000-0 (sc)
ISBN-000-0-0000-0000-0 (hc)
ISBN-000-0-0000-0000-0 (eBook)

All rights reserved. No part of this book may be reproduced or transmitted in any form or by any means, electronic or mechanical, including photocopying, recording, or by any information storage and retrieval system, without permission in writing from the copyright owner.

The views expressed in this work are solely those of the author and do not necessarily reflect the views of the publisher, and the publisher hereby disclaims any responsibility for them.

Matchstick Literary
1-888-306-8885
orders@matchliterary.com

DEDICATED:

My mom Deboroah Bewley who provided years of educational expertise for the production of this book.

Prologue

The only explanation for the origin of the universe, Earth and life itself found in the textbook is "evolution" and "millions of years". Evolution is taught as an undeniable fact that no one should question.

This work will discuss the scientific flaws and other inaccuracies that are present in many school textbooks around the world.

Of course, not all of these misconceptions will be found in every textbook. However, these concepts are held by the scientific community, even if they are not in your textbook. Although these errors exist in all forms of media, for the sake of simplicity, this work will refer to all of these mediums collectively as the textbook.

This work is designed to provide as much information as possible to help students, parents, and teachers better understand what the textbook teaches and why it is wrong.

These errors are not just present in your textbooks, but children's books and every form of media. These inaccuracies are perpetuated without evidence from even before children can read.

Put another way, if your math book said that 2+2=5, would you correct this error or just accept it because that is what the textbook says?

In short: You are the Watchman for your child and must stand up and blow the trumpet.

Table of Contents

Introduction: Teaching by Reason 1
    1) Why Do We Wash Our Hands 3
    2) What Is In Our Drinking Water 4
    3) A New Model of the Universe 5
1. Evolutionary Ideology 9
    1.1. Darwin's Job 11
    1.2. Definitions 13
    1.3. Scientists of the Past 15
2. Ancient History Part I 21
    2.1. Tectonic Plates 23
    2.2. The Geologic Column 24
    2.3. Fossilization 26
    2.4. Oil and Diamonds 27
3. Ancient History Part II 27
    3.1. Good Geologic Clocks 33
    3.2. Bad Geologic Clocks 35
    3.3. Dinosaurs 39
    3.4. The First Cells 43
    3.5. The Ages of Humanity 46
4. Radioactive Dating 51
    4.1. Why Dating Methods Fail 53
    4.2. How Dating Methods Fail 56
    4.3. When to Use Dating Methods 57
5. Forming Elements 63
    5.1. Formation of Atoms 65
    5.2. Formation of the Universe 67
6. Genetic Changes 71
    6.1. Darwin's Finches 73
    6.2. Genetic Variability 74
    6.3. Antibiotic Resistance 80
    6.4. Variations and Hybrids 81
    6.5. Variations in Humans 82
7. Transitional Forms 87
    7.1. Transitional Forms 89
    7.2. The History of Animals 91
    7.3. Dinosaurs and Birds 93
    7.4. The History of Humans 96
    7.5. Variation in Fossils 99
    7.6. Common Bones 101
8. Vestigial Organs. 107
    8.1. Vestigial Organs in Animals. 109
    8.2. Vestigial Organs in Humans 110

  8.3. Junk DNA 112
  8.4. Non-Existent Organs 113
9. Starlight 119
  9.1. Measuring Stars 121
  9.2. Locating Stars 123
  9.3. Birth of Stars 125
  9.4. Death of Stars 109
  9.5. Forming Elements 127

Epilogue XXX
Glossary XXX

Operation: Battleground Textbook

Introduction
Teaching by Reason

What Does History Tell Us?

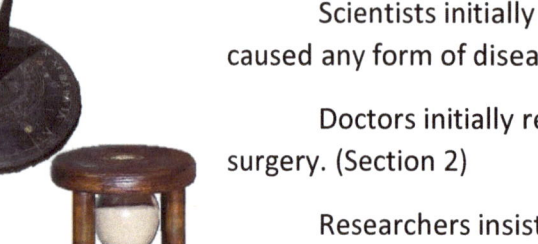

Scientists initially scoffed at the very idea that microbes caused any form of disease. (Section 1)

Doctors initially refused to wash their hands before surgery. (Section 2)

Researchers insisted that the Earth was at the center of the universe. (Section 3)

Section 1
Why Do We Wash Our Hands?

We all know that it is a good idea to wash your hands, but did you ever think to ask why this exercise came into practice?

Ignaz Semmelweis (Figure I.1.1) was appointed as an assistant at the Obstetric Clinic in Vienna. There, doctors frequently practiced with cadavers (bodies used for medical research and dissection) before delivering babies in the maternity ward (Figure I.1.2).

Women would very often die. The mortality rate was quite high because of working with dead bodies and the diseases present.

One day, Dr. Semmelweis made a radical suggestion: *Doctors should wash their hands with soap and chlorine after dealing with cadavers* (Figure I.1.3).

Fig. I.1.1 Ignaz Semmelweis

Fig. I.1.2 An obstetrician.

The practicing doctors were appalled at this suggestion because it seemed to imply that they were passing on illnesses to women.

However, despite opposition, the doctors that began implementing this practice had a much lower mortality rate.

His idea saved countless lives and was supported by the governing bodies until a conference of German physicians rejected the handwashing doctrine.

Fig. I.1.3 Washing hands with chlorine.

He was publicly berated and made many influential enemies.

It was not long before doctors began giving up the practice of handwashing. Opposition from his colleagues caused Semmelweis to lose his job.

He kept trying to convince doctors but had varying levels of success on his own.

Today, Dr. Semmelweis has been vindicated, but many doctors and government officials refused to listen to him.

Just because the majority of scientists do not believe in a new or revolutionary idea does not make it untrue.

Just because the scientists do not understand everything about the science does not make it false.

It is the responsibility of those in academia to teach what is scientifically accurate above all else.

Section 2
What Is In Our Drinking Water?

Miasma Theory is a relatively recent explanation for diseases and illness. Miasma Theory postulated that poisonous or bad air was the cause of all disease. The Black Plague was thought to be spread through the air. Doctors even wore special masks (Figure I.2.1) in response to the Miasma Theory.

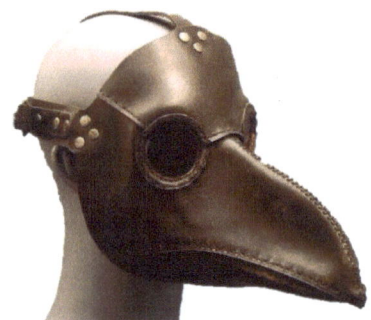

Fig. I.2.1 A doctor's mask for Miasma Theory

In the 17th Century, the first microscopes (Figure I.2.2) allowed scientists to find microscopic organisms. Some suggested that these organisms were the cause of some types of disease.

Fig. I.2.2 An early microscope

However, many in the scientific community rejected the notion of germs and stuck to their belief in bad air.

The supporters of Germ Theory were determined to prove Miasma Theory wrong.

Dr. John Snow (Figure I.2.3) was one such doctor. After examining all of the data available, Snow was convinced that Cholera was caused by an infected water supply (Figure I.2.4).

Fig. I.2.3 Dr. John Snow

When Cholera came to the Soho region, he found that most of the deceased had lived near a water pump. Since Germ Theory would not be accepted, he suggested that it was a toxin in the water and not the air that caused the outbreak.

Fig. I.2.4 Filthy water

City officials still refused to accept his idea even after shutting off the pump caused the number of cases to drop substantially.

Today, Dr. Snow has been vindicated, but many scientists and government officials refused to listen to him.

Just because the majority of scientists do not believe in a new or revolutionary idea does not make it untrue.

Just because the scientists do not understand everything about the science does not make it false.

It is the responsibility of those in academia to teach what is scientifically accurate above all else.

Section 3
A New Model of the Universe

Fig. I.4.2 Galileo Galilei

Nicholas Copernicus (Figure I.4.1) was famous for putting forth the idea that the Earth is not at the center of the universe (heliocentric). The work of Copernicus inspired Galileo Galilei (Figure I.4.2) to do his research that only furthered these findings. As such, our knowledge of the universe was significantly expanded.

As a prominent defender of heliocentrism, Galileo was scorned by the astronomers of his time who were staunch advocates of geocentrism. This taught that the Earth was at the center of the universe.

Fig. I.4.1 Nicholas Copernicus

He, like Copernicus, proposed that the sun was at the center of the solar system (Figure I.4.3).

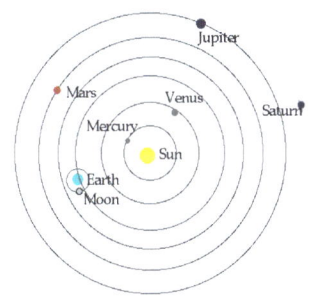

Fig. I.4.3 A geocentric universe

To discredit Christianity, the evolutionists have essentially rewritten history to retell the so-called Galileo Affair. The most common telling is that Galileo disrupted the views of the Catholic Church that held that the Earth was at the center of the cosmos (Figure I.4.4).

There is no historical evidence to support this intentionally corrupted history.

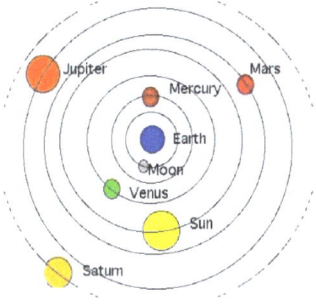

Fig. I.4.4 A heliocentric

In reality, the church was very accepting of Galileo, but it was the scientific community that ignored the results and fought to get the church to call Galileo a heretic.

The scientists at the time fought to keep the teaching that the Earth was at the center of the universe and this made astronomy stagnant.

Today, both Copernicus and Galileo have been vindicated, but many scientists and officials refused to listen to him.

Just because the majority of scientists do not believe in a new or revolutionary idea does not make it untrue.

Just because the scientists do not understand everything about the science does not make it false.

It is the responsibility of those in academia to teach what is scientifically accurate above all else.

In Conclusion:

The mere idea that one should wash their hands after performing certain tasks was ridiculed in its time for fighting against what the established scientists beleived. (Section 1)

Likewise, teaching that there might be harmful organisms that cause disease, living within the water supply was fought against to the bitter end. (Section 2)

When scientists proposed that the Earth was not at the center of the universe, the scientific community fought back. (Section 3)

Questions for Further Discussion:

1. Why did scientists not listen to Dr. Semmelweis? (Section 1)
2. Why did scientists not listen to Dr. John Snow? (Section 2)
3. Why did scientists not listen to Copernicus or Galileo? (Section 3)
4. What happens when scientists stop debating and asking questions?
5. What other scientific phenomena do you think have been falsified?

NOTES:

CHAPTER 1
Evolutionary Ideology

Chapter 1: Evolutionary Ideology

What Does the Textbook Teach?

The textbook will teach that Darwin was a respected scientist. The textbook also teaches that during his journey aboard the HMS Beagle, he truly recognized the history of the world and the common ancestry of all organisms. (Section 1)

It will also teach that we should accept evolution because of how a theory is defined. (Section 2)

Finally, it will teach you that no rational person, let alone a scientific intellectual would even consider that the Earth is not millions of years old. (Section 3)

Section 1
DARWIN'S JOB:

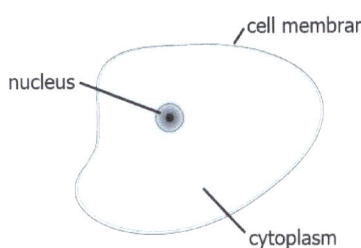

Fig. 1.1.1 Charles Darwin

The textbook will say that Charles Darwin (Figure 1.1.1) was a naturalist aboard the HMS Beagle when he went to the Galapagos Islands.

In reality, he was a theology student that could not get a job with this degree. Darwin did study multiple degree programs but never finished any of them except for theology.

He eventually got a job as the captain's companion. In those days, the captain was not allowed to interact with the crew on a personal level. Darwin's job was to keep the captain company and to provide someone to talk to.

There is no evidence to suggest that his travels were for his research or that he was brought on board because of his credentials.

Many of the things that we are taught about scientists in the past are wrong. These myths are often propped up and the truth ignored so that we will not scrutinize their findings or credentials. If we did look more closely, we would be less likely to believe what they taught.

Another overlooked fact is that in Darwin's time, our knowledge of biology, especially cell biology, was very limited.

This limited knowledge of the cell led Darwin and the scientists at the time to vastly overestimate the possibility of evolution.

Scientists at the time thought that the cell was little more than a membrane with genetic material (Figure 1.1.2). We now know that the cell is infinitely more complex than anything that the scientists at the time could have ever imagined (Figure 1.1.3).

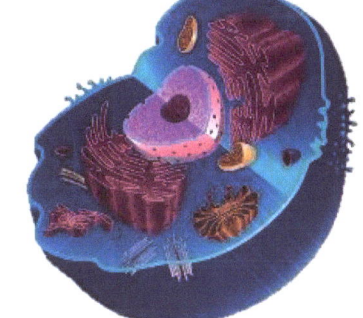

For comparison, those in Darwin's time thought that a cell had the complexity of a hut, but we now know that it is as complex as an entire city (Figure 1.1.4).

Fig. 1.1.2 An early cell model

Fig. 1.1.3 A modern cell model

Fig. 1.1.4 A hut and a city

The limited knowledge of Darwin and his contemporaries concerning microbiology hindered how they saw the world. Because scientists thought that the cell and therefore any organism was simple, it seemed obvious that all organisms could have had a common ancestor just a few million years ago. With present microbiology, it is much more farfetched to hold this assessment.

Section 2
Definitions:

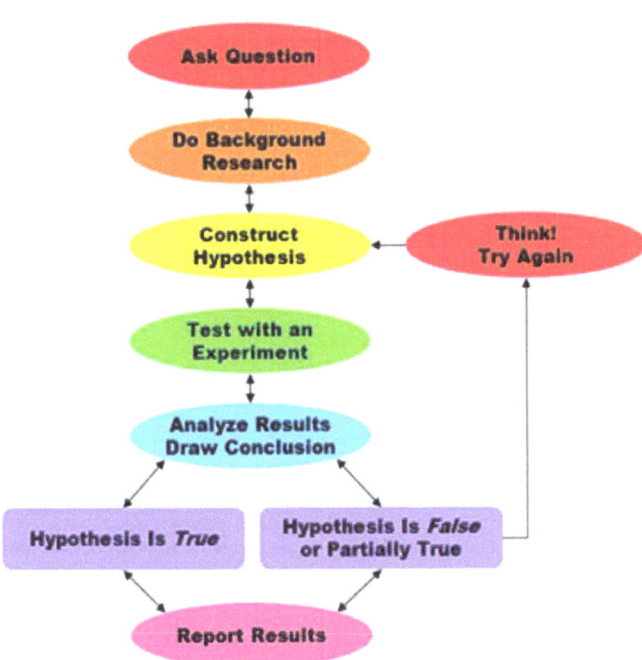

Fig. 1.2.1 The Scientific Method

The textbook will say that evolution is a theory, just like gravity. First, we do need to define a few terms.

A hypothesis is an idea that has yet to be tested. One of the first steps of the Scientific Method is to hypothesize (Figure 1.2.1). When a hypothesis is disproven, it is normally discarded. However, because so many scientists are such adamant supporters, they will continue to advance the concept of evolution regardless of the evidence that comes to light.

A theory is a hypothesis that has been tested and has held up to scrutiny. A theory can still be disproven if more evidence comes to light.

A law is a theory that has been proven true beyond all doubt. This is often reserved for mathematical proofs such as $E=MC^2$.

Now the question is: Which of these definitions best fits evolution?

Throughout this work, we will see that evolution does not hold up to scientific scrutiny. Thus, the argument that evolution is a theory, just like gravity is nonsense.

Let us analyze gravity to see where it falls based on the scientific method:
1. Ask a Question:
Do all objects fall to the Earth?
2. Construct a Hypothesis:
I predict that objects will fall to the Earth regardless of what the object is.
3. Test With Experiment:
I dropped 250 different objects to the ground from 1.8 meters.
4. Analyze Results:
Every single object fell to the ground, although some fell faster than others.
5. Find Solution:
Therefore, all objects will fall to the ground.

Because of this and many repeated experiments, scientists found that gravity always causes an object to fall to the ground. This became known as The Law of Gravity.

Now, the question is why do objects fall to the ground? The reason why the objects fall is the Theory of Gravity.

A proposed explanation for why objects fall to the ground would be a Hypothesis of Gravity.

The rest of this work will show how the evidence used to support the evolution hypothesis* lacks empirical data.

Two more important definitions need to be established: experimental science and historical science.

Experimental science refers to that which can be tested and observed via the scientific method. This is the branch of science that has brought forth spectacular achievements in the field of science such as space shuttles and medicine (Figure 1.2.2).

Fig. 1.2.2 Space shuttles and medicine

Fig. 1.2.3 Historical science

Historical science depends on deductive reasoning to put together clues to understand the past. This branch is much more prone to bias. Historical science is the science behind forensics (crime scene investigation) (Figure 1.2.3).

Once the forensic scientist observes all of the facts, he will attempt to try to put a narrative together to explain how the crime occurred. Naturally, if the scientist is certain that there was a murder before looking at the evidence, even if there is no evidence of a murder, he may still conclude that a murder took place.

The same is true for evolution. If a scientist is certain that the Earth is millions of years old that is how he will interpret the data.

There is not a single component of any branch of science (save for evolutionary biology) that has made any contributions with the millions of years of evolution hypothesis.

---

*This work will refer to evolution as a hypothesis rather than a theory since it does not meet the qualifications for the latter.

Section 3
Scientists of the Past:

There are countless occurrences, across history, in which the majority of scientists believed something contrary to what was observed. Two prime examples of this bias are the flat earth and geocentrism (the hypothesis that the Earth is at the center of the universe). Many of the scientists who fought against the establishment to get these lies out of academia were scorned by their peers. Often, they were not vindicated until after their death.

The textbook will say that the majority of scientists throughout history have believed in evolution. Even if this were true, that does not make evolution true.

In contrast, here are just a few of the famous and ground-breaking scientists that believed in a young Earth:

Fig. 1.3.1
Francis Bacon

Fig. 1.3.2
Carl Linnaeus

- Francis Bacon (Figure 1.3.1) developed the scientific method that we still use today.
- Dr. Raymond Damadian created the MRI that scans the human brain.
- Dr. John Baumgardner founded the modern study of catastrophic plate tectonics.
- Carl Linnaeus (Figure 1.3.2) developed the modern classification system.
- Dr. John Sanford developed the gene gun.
- Werner Von Braun was the founder of rocket science that helped to get a man to the moon.
- Sir Isaac Newton (Figure 1.3.3) was the founder of modern-day physics.
- Galileo Galilei's (Figure 1.3.4) observations furthered our understanding of the universe and our planet's place in the solar system.
- Sir David Brewster's work led to a better understanding of how the lens in the eye worked.
- Louis Pasteur (Figure 1.3.5) developed a process to kill microorganisms to keep milk fresh longer.
- Michael Faraday's work in metallurgy helped lead to the discovery of how electromagnetism causes motion.
- Joseph Lister (Figure 1.3.6) helped bring about the practice of sterilizing clothing and equipment before surgery.

Fig. 1.3.3
Sir. Isaac Newton

Fig. 1.3.4
Galileo Galilei

Fig. 1.3.5
Louis Pasteur

Fig. 1.3.6
Joseph Lister

Fig. 1.3.7
George Washington Carver

- James Clerk Maxwell showed that magnetism, electricity, and light were simply different manifestations of the same fundamental laws.
- George Washington Carver (Figure 1.3.7) was a chemist who found countless uses for the peanut.

These are just a few examples of the most brilliant scientists in history and founders of their respective fields that believed in a young Earth and did not accept the millions of years hypothesis which is vital to evolution.

IN CONCLUSION:

Although Darwin had no qualifications as a scientist, his findings are still taken as fact. (Section 1)

Furthermore, the actual definition of the word "theory" in regards to evolution does not fit what we observe via empirical science. (Section 2)

Finally, many renowned scientists across history believed in a young Earth. (Section 3)

Questions for Further Discussion:

1. Why would people boast Darwin as a scientist when he did not have any formal training or authority? (Section 1)
2. Why would evolution be called a theory when it only fits the definition of a hypothesis? (Section 2)
3. Why would the textbook tell us that all scientists throughout history believed in millions of years when that is factually untrue? (Section 3)
4. What other scientific phenomena do you think have been falsified?

Additional Reading

Secular Sources:

"Historical Perspective, Darwin and Evolution." *Cell Theory*, University of Miami, www.bio.miami.edu/tom/courses/bil160/bil160goods/03_darwin.html.

Lennox, James. "Darwinism." *Stanford Encyclopedia of Philosophy*, Stanford University, 26 May 2015, plato.stanford.edu/entries/darwinism/.

Religious Sources:

"Scientists of the Past Who Believed in a Creator." *Creation.com | Creation Ministries International*, Creation Ministries International, creation.com/scientists-of-the-past-who-believed-in-a-creator.

NOTES:

Chapter 9: Starlight

NOTES:

CHAPTER 2
Ancient History Part I

Chapter 2: Ancient History Part I

What Does the Textbook Teach?

The textbook is going to teach that the Earth's surface has been changing slowly over millions of years. (Section 1)

It will also tell you that scientists have evidence of when creatures around the world lived based on fossils that formed over millions of years. (Section 2)

The textbook will then describe that fossils are not the only things that take millions of years to form. (Section 3 & Section 4)

Section 1
Pangea:

The textbook will say that at some point in Earth's past, there was only one continent called Pangea (Figure 2.1.1). There is no empirical or observable science to support this.

However, the textbook will say that the best support of this is how well South America and Africa fit together. The textbook will not tell you that the dimensions of the two continents are radically shifted in between the pictures (Figure 2.1.2). The reason that they must change the dimensions between the pictures is that the dimensions do not fit what the scientists expect to see.

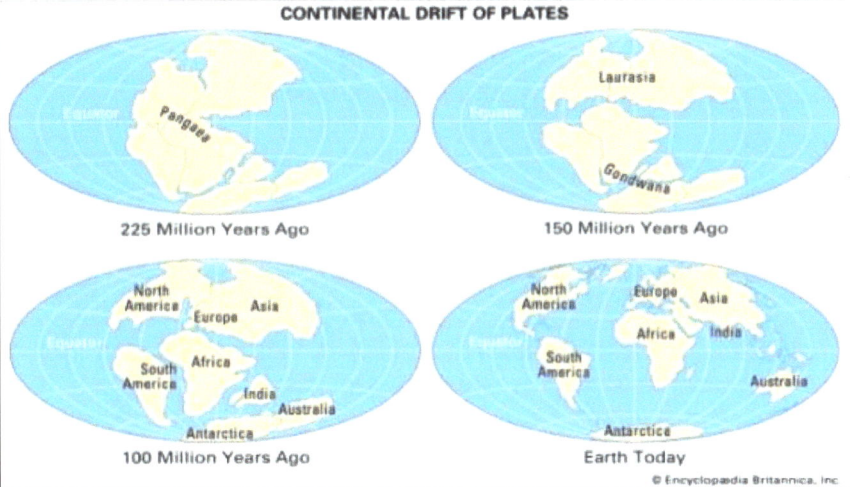

Fig. 2.1.1 The movement of the continents.

Furthermore, there are still other discrepancies:

Measurement	Legnth	Variation	Height	Variation
Africa 1	61	148.78%	53	88.33%
South America 1	41		60	
Africa 2	60	150.00%	55	85.94%
South America 2	40		64	
Africa 3	61	145.24%	56	100.00%
South America 3	42		56	
Africa 4	57	150.00%	54	103.85%
South America 4	38		52	

- Antarctica drifts only a fraction of the distance that South America and Australia drift.
- Most of southern Europe appears between stages three and four.
- The Middle East appears and radically shifts position between each stage.
- Most of Central America just appears between the third and fourth stages.

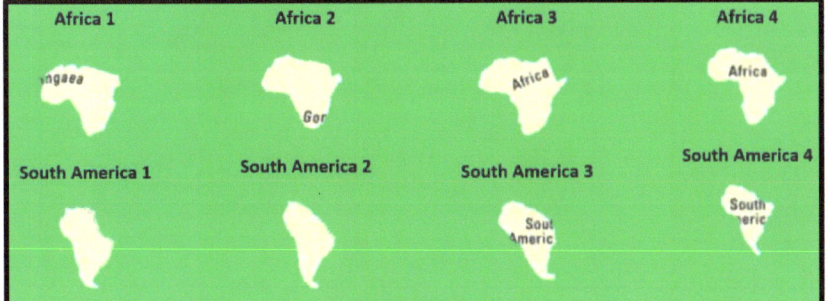

Fig. 2.1.2 The shift in dimensions.

The reason that the textbook will ingrain the idea of Pangea is that it adds credence to saying that the Earth is millions of years old.

Section 2
The Geologic Column:

Fig. 2.2.1 Charles Lyell

In the 19th Century, Charles Lyell (Figure 2.2.1) published Principles of Geology. This work claimed not only that the Earth was millions of years old, but that there was a so-called geologic column (Figure 2.2.2).

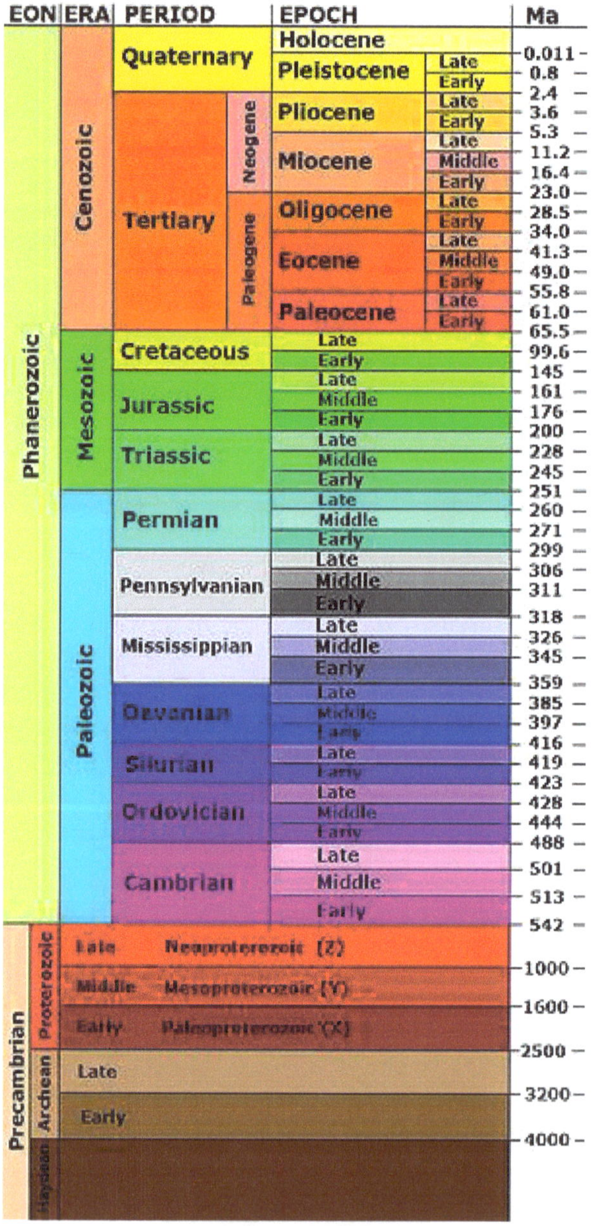

The textbook will say that when an organism dies if it fossilizes, scientists expect a methodological placement of the organism.

The textbook will further describe that over time, dirt and other sediments will form on top of it. Over millions of years, other organisms will fossilize and be buried in the next layer up. If this pattern continues, you would, in theory, be able to see this geologic column.

Lyell, gave each period a name, age, and index fossil. These ages were put forth before any dating method had been discovered. (This will be further examined in Chapter 4.) Dating fossils would have been impossible if the geologic column had not been constructed first.

Lyell created his geologic column without having a dating method. Despite the acceptance of Lyell's claims, there are only a handful of places where the fossils are seen in the right order.

Even if these fossils are found in the right order, there are still problems.

Introducing the coelacanth (Figure 2.2.3). This is a fish has been seen off of the coast of Madagascar, in spite of the "evidence" that it should have died out millions of years ago.

It is found very early in the fossil record and thus the textbook will conclude that this creature appeared very early on in Earth's history.

Fig. 2.2.2 The geologic column

Fig. 2.2.3 The Coelocanth

However, since this fish is alive today, they should be present throughout the fossil record. Despite how the real world evidence requires that these fish be present throughout the geologic column, they are not fossilized in the upper layers of the column or in the same layer as humans.

The next major misrepresentation of scientific fact in the textbook is in regards to dinosaurs.

The textbook will say that dinosaurs evolved into birds. However, we find birds and dinosaurs in the same layer.

Scientists have found evidence to suggest that dinosaurs ate birds (Figure 2.2.4). Contrasting the evolution hypothesis, an organism cannot evolve into something that already exists.

Fig. 2.2.4 A Microraptor attacking a bird.

Furthermore, fossils have been found in the wrong layer far more often than the evolutionist wishes to admit. Every time that a fossil is found in the wrong place, it should motivate scientists to rethink Earth history. This update to the evolutionary world view never occurs.

The reason that the geologic column needs to be updated is that it is based on false and unverifiable assumptions. Paleontologists question the order of the geologic column rather than questioning whether or not the geologic column does exist.

Fig. 2.2.5 A poly-strata tree

Next, we have poly-strata fossils (Figure 2.2.5). These fossils are mostly trees that poke through multiple layers of strata. Since the textbook will tell you that each layer took millions of years to form, these fossils should not exist.

Trees, like any other living organism, require nutrients to live. If they do not have them, the organism will die. If this tree stood by for millions of years while the layers formed around it, it would have rotted away in the meantime.

Section 3
Fossilization:

One of the vital components for the geologic column is gradual fossilization (Figure 2.3.1). However, the same problem with poly-strata fossils comes up with every single organism.

Fig. 2.3.1 A fossil

Post haste burial is the key to fossilization. An organism can only fossilize if it is covered in sediment quickly to prevent decomposition and oxidation. Contrary to the description of fossilization that is found in most textbooks, the burial of the organism must occur immediately for a fossil to form.

Fig. 2.3.2 A jellyfish fossil

The textbook will say that soft tissue rarely ever gets fossilized, but this is untrue.

Jellyfish do not have any bones and are all soft tissue, but they are very readily fossilized (Figure 2.3.2).

Fig. 2.3.3 Soft tissue from a Tyrannosaurus femur

Another problematic discovery is soft tissue in bones. While transporting a tyrannosaurus femur, paleontologists accidentally dropped a bone and broke it in half. This was initially thought to be a tragedy until they took a closer look. The scientists found soft tissue and blood cells inside (Figure 2.3.3). Since dinosaur bones should be millions of years old, any piece of soft tissue could not have survived for epochs of time.

To further complicate the issue, scientists have made fossils in a lab that are indistinguishable from fossils that form in nature. It only logically follows that since fossils do not take millions of years to form, the ages given to the geologic column must be re-evaluated.

Fig. 2.3.4 Fossil of a fish eating another fish

Fossilization occurs very quickly because the soft tissue is found. Scientists have also found fossils of animals eating each other (Figure 2.3.4) and giving birth (Figure 2.3.5). Both are incredibly fast. Therefore, the fossilization did not occur over millions of years.

Fig. 2.3.5 Fossil of an animal giving birth

Section 4
Oil and Diamonds:

Fossils are not the only items that the textbook will say take millions of years to form. Purportedly, oil (Figure 2.4.1) and diamonds (Figure 2.4.2) also take millions of years to form.

Fig. 2.4.1 An oil rig collecting "ancient" oil.

However, since both have been made in a lab, it is nonsensical to say that they took millions of years to form. These scientists claim that they did in a short time what took nature millions of years to do.

There is no evidence that the Earth is millions of years old based on what develops out in nature.

Fig. 2.4.2 A diamond that formed very quickly.

Chapter 2: Ancient History Part I

IN CONCLUSION:

The maps created to show how Pangea eventually broke into the continents that we see today are doctored to force them to fit in with the assumptions inherent in millions of years. (Section 1)

The timeline of Pangea is not based in science and neither is the geologic column. (Section 1 & Section 2)

There are a vast number of organisms that are fossilized with and without bones and because of this, the fossilization narrative given by the evolutionists is severely flawed. (Section 2)

Since soft tissue does not last over millions of years, perhaps the tissue or fossils are not as old as the evolutionists claim. (Section 3)

Diamonds and oil can arise very quickly and do not take millions of years to form. (Section 4)

Questions for Further Discussion:

1. Why was the map for Earth's history in regards to Pangea changed? (Section 1)
2. Why would scientists force the idea of Pangea? (Section 1)
3. Is there any evidence supporting the geologic column? (Section 2)
4. If soft-bodied creatures are fossilized, what does that mean about fossilization? (Section 3)
5. If soft tissue is still present in bones, what does that mean about fossilization? (Section 3)
6. Why would scientists claim that fossils, oil, or diamonds take millions of years to form? (Section 3 & Section 4)
7. Is there another hypothesis that will explain these phenomena?

Additional Reading

Secular Sources:

Britannica, The Editors of Encyclopaedia. "Pangea." *Encyclopædia Britannica*, Encyclopædia Britannica, Inc., 3 Dec. 2018, www.britannica.com/place/Pangea/media/441211/172046.

Britannica, The Editors of Encyclopaedia. "Pangea." *Encyclopædia Britannica*, Encyclopædia Britannica, Inc., 3 Dec. 2018, www.britannica.com/place/Pangea.

"Fossil Jellyfish." *Ocean Portal | Smithsonian*, Smithsonian's National Museum of Natural History, 18 May 2018, ocean.si.edu/ocean-life/invertebrates/fossil-jellyfish.

"Geologic Column." *The U.S. Naval Academy*, www.usna.edu/Users/oceano/pguth/md_help/geology_course/geologic_column.htm.

Li, Yebo, and Caxia Wan. "Algae for Biofuels." *Safe Operation of Utility Type Vehicles (UTVs)*, Ohio State University, 11 Apr. 2011, ohioline.osu.edu/factsheet/AEX-651-11.

Peake, Tracey. "Soft Tissue Analysis: Jurassic Ichthyosaur Was Warm-Blooded, Had Blubber and Camouflage." *NC State News Bigger Than Football Study Shows Sports Can Help Communities Recover From Disaster Comments*, North Carolina State University, 5 Dec. 2018, news.ncsu.edu/2018/12/ichthyosaur-blubber/.

"Pure CarbonMan Made Diamonds." *Moissanite Engagement Rings*, Diamond Nexus, www.diamondnexus.com/pure-carbon-man-made-diamonds.html.

"Uniformitarianism: Charles Lyell." *Reproductive Isolation*, Berkeley University, evolution.berkeley.edu/evolibrary/article/history_12.

Chapter 2: Ancient History Part I

Additional Reading

Religious Sources:

University, Southern Adventist. "GC A3 Geologic Column--Relative Dating." *Southern Adventist University*, Southern Adventist University, www.southern.edu/academics/academic-sites/faithandscience/Origins-Curriculum-Resources/i-3-GC-A3-Geologic-Column--Relative-Dating.html.

University, Southern Adventist. "GC B3 Formation of Layers." *Southern Adventist University*, Southern Adventist University, www.southern.edu/academics/academic-sites/faithandscience/Origins-Curriculum-Resources/i-3-GC-B3-Formation-of-Layers.html.

NOTES:

Chapter 9: Starlight

NOTES:

Operation: Battleground Textbook

CHAPTER 3
Ancient History Part II

Chapter 3: Ancient History Part II

What Does the Textbook Teach?

The textbook will say that the Earth is millions of years old and that there are several scientific measurements to prove this. (Section 1)

Scientists will claim that the Earth is millions of years old based on flawed measurements. (Section 2)

The textbook will say that dinosaurs lived millions of years ago. (Section 3)

The textbook will also teach that life arose from non-life. (Section 4)

The textbook will describe that there was a stone age, iron age, and bronze age. (Section 5)

Section 1
Good Geologic Clocks:

Evolution via millions of years is based on one central assumption, gradualism. Gradualism is the philosophy that everything that we see today has always been occurring at the same rate.

This is in stark contrast to catastrophism, which states that there are events in Earth's past that radically changed Earth's history in a very short time.

The world view of the observer will change how he sees the evidence.

The moon and the ocean are two good measurements to determine Earth's relative age.

Fig. 3.1.1 The Earth and the Moon

The moon has a strong effect on Earth. The moon has to be far enough away from the Earth to not get pulled back in and crash down to the surface, but close enough so that it does not just float away. Naturally, no distance is just right and the moon is drifting away from the Earth at a rate of about 4 centimeters per year.

Because of modern astrophysical observations, we know that the moon is about 240,000 miles from the Earth on average. Let us compare what we observe to what millions of years would predict as opposed to what a much younger Earth shows.

If the Earth-moon system is 4.5 billion years old, and assuming gradualism, the moon should have drifted 11.2 million miles or 46 times the distance between the Earth and the moon.

If we assume that the Earth-Moon system is much younger, such as 10,000 years we get a very different number.

At 4 centimeters per year, this puts the total distance that the moon has drifted at 0.249 miles.

In this instance, like many others, a world view of millions of years does not coincide with observed evidence.

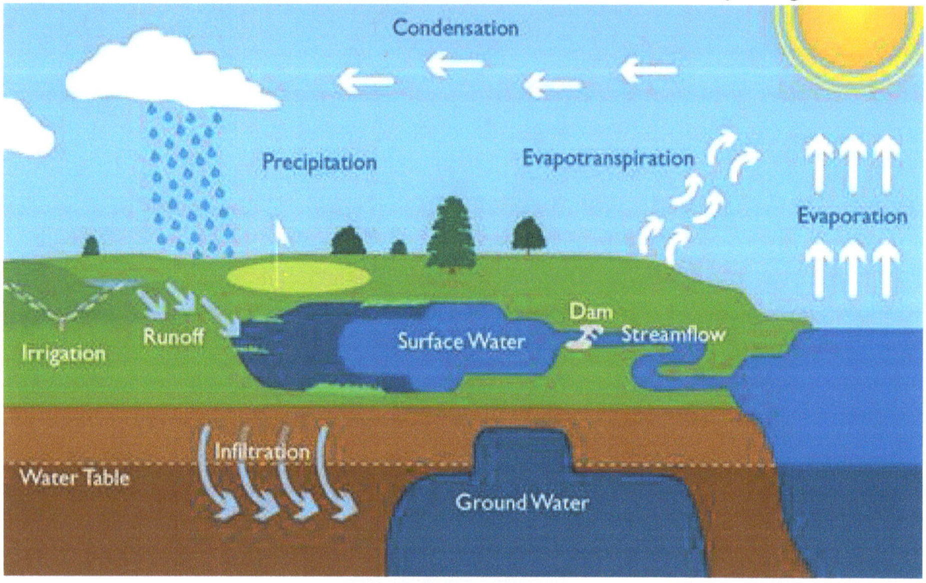

Fig. 3.1.2 The Water Cycle

Another geologic clock for calculating the age of the Earth is the salinity (saltiness) of the oceans.

As rain and other natural forces push sediments into the ocean, it becomes saltier (Figure 3.1.2). When evaporation occurs, the salt remains, but the water returns to the atmosphere.

Therefore, if we calculate the average rate today as to how quickly the salinity of the oceans is increasing, we can (assuming gradualism) calculate how old the Earth is.

Scientists Dr. Steve Austin and Dr. Russell Humphreys, originally calculated the age of the oceans to be a maximum of 62 million years old. The only problem is that Austin and Humphreys assumed the lowest input of sediments possible.

Scientists now know that the salinity of the oceans is increasing far faster than these two had originally proposed (catastrophism). Austin and Humphreys expected the sediments to be .01-10% from runoff, mostly rivers. However, after studying radium in coastal waters, scientists now believe that the runoff may be as much as 40% from river flow.

This means that the river flow may be much faster than Austin and Humphreys originally thought. If true, the oceans have been increasing in salinity faster and thus to reach the concentration that we see today, the Earth could not possibly be millions of years old.

The problem is that Austin and Humphreys began with an assumption that was altered by their world view.

Therefore, since using gradualism contradicts what we see in nature, perhaps we need to consider that large portions of the Earth were shaped by catastrophic forces.

Floods, volcanos, earthquakes, storms, and many other natural phenomena (Figure 3.1.4) can change an entire landscape overnight. Sometimes the changes can occur in just a few minutes. Instances of large environments being changed drastically by a single event are quite common throughout history.

Since it is an unquestionable fact that massive changes occur very quickly, the entire notion of gradualism should not be trusted as the primary method to date the Earth or its history.

Fig. 3.1.4 Natural disasters

In addition to the moon and ocean salinity, let's consider a stellar example to calculate the earth's age. Comets are extremely fragile and they break apart relatively quickly, especially if they pass by a star.

Based on calculations by astronomers, and assuming millions of years, there should be no comets left after millions of years.

Evolutionists get around this by saying that the comets formed in the Oort cloud, but there is no physical evidence that this structure exists.

Chapter 3: Ancient History Part II

Section 2

Bad Geologic Clocks

Just as there are good ways to measure the age of the Earth, there are also very bad ones.

One of the more common clocks used by scientists is ice cores (Figure 3.2.1). Ice cores are pieces of ice that have been pulled out of the ground that can be several feet in length. Since the substances are frozen, the ice cores show scientists the atmospheric and chemical composition of whatever was present at the time that the ice was laid down.

Fig. 3.2.1 An Ice Core

Each ice core alternates between dark and light stripes. Supposedly these stripes represent summer and winter.

The best explanation for why ice cores are not accurate geological clocks is the story of the Lost Squadron of WWII.

In November 1942 for Lieutenant Pritchard, the pilot, and Petty Officer Bottoms, the radio operator, were flying through a terrible storm. It got so bad that Pritchard had to make a crash landing in Greenland (Figure 3.2.2).

Fig. 3.2.2 Greenland

The question was a amphibious Duck (Figure

Fig. 3.2.3 Amphibious Grumman Duck

plane in single-engine Grumman 3.2.3).

Seventy years later, a crew was sent to find it. Even though the crew was searching for a plane in the middle of Greenland, they did not think it would be that hard. The team brought a metal detector and expected the craft to have survived under only a few feet of snow.

After a very long time, they found the craft, not under a few feet of snow, but 1.16 m or 38 ft down under ice.

The European Geosciences Union (EGU) found a 3.2 Km ice core that supposedly goes back 800,000 years. Remember that this ice core is used as a geological clock. Utilizing simple ratios, this means that if we assume long ages, the airplane crashed more than 2,000 years ago.

The plane crashed in 1942, thus the ice it was found under should be no deeper than 0.28 m or 10.92 in.

It seems more likely that the stripes just represent alternating warm and cold weather. The weather can change very drastically over just one week.

Therefore the ice cores that are used to show what happened on the Earth millions of years ago only show what has happened relatively recently.

Another often used geologic clock are the rings of trees. Supposedly, the tree will get a new ring each year of its life (Figure 3.2.4). However, that is not always true.

The tree will get a new ring in relation to how much water there is available.

Unlike the distance between the Earth and the moon or the salinity of the oceans, these tests are unreliable, yet they are two tests that scientists use to test how old the Earth is. The only problem is that they are based on faulty assumptions.

The empirical evidence does not support that the Earth is even millions of years old.

Fig. 3.2.4 The growth rings of a tree

Chapter 3: Ancient History Part II

Section 3
Dinosaurs:

The textbook will say that dinosaurs died out millions of years ago. However, did you ever stop to think why we are so certain that they died out millions of years ago?

As discussed in the previous chapter, the main method of dating the fossil of an organism is by determining which layer it came from. Since this is unscientific and has no facts supporting it, we need to question this reasoning. This should lead scientists to conclude that the dinosaurs lived much closer to the present day than previously theorized.

Fig. 3.3.1 The Platypus

The evolutionist reasons that the now extinct dinosaurs were unique and exotic. Because they are not alive and are truly bizarre, humanity must have never seen them.

Fig. 3.3.2 The Chimera of Greek Mythology

Consider the platypus (Figure 3.3.1). It is a creature with such strange features that when it was first reported, many thought that it was a hoax.

Such as mashup of creatures is something that would be expected in the world of mythology (Figure 3.3.2). However, skeptics were eventually proven wrong when live platypuses were captured. We see many other eccentric creatures even today.

Fig. 3.3.3 The Tufted Deer

Fig. 3.3.4 The Blob Fish

- The Tufted Deer (Figure 3.3.3) is a deer that has fang-like teeth.
- The Blob Fish (Figure 3.3.4)
- The Japanese Spider Crab (Figure 3.3.5) has a leg length of 18 feet from claw to claw.

Fig. 3.3.6 The Dodo

Fig. 3.3.5 The Japanese Spider Crab

Also, consider the dodo (Figure 3.3.6). It, like the dinosaurs, is extinct. No one is going to deny that there are a vast number of creatures of just about every shape and size around the world. Why is it so farfetched to believe that there were creatures that were alive in the past that are now extinct that were truly spectacular?

Therefore, it is logically unsound to say that just because there are no dinosaurs alive today that they lived millions of years ago.

We have looked at the fact that there were exotic creatures in both the past and the present and the dating methods are unreliable. Now the question remains: do we see dinosaurs in any historical account other than paleontology?

Fig. 3.3.7 Babylonian Gate 600 BC

The word dinosaur did not exist until 1842 and means "Terrible Lizard". You will not find the word dinosaur in any historical record, although they may have been called dragons in the past.

Fig. 3.3.9 Ica Stones of Peru

These amazing "dragons" are depicted in artwork all over the world and across history.

Fig. 3.3.8 Cambodian Temple

Here are just a few examples of the hundreds of artistic depictions all over the world. Far too many are shown in paintings (Figure 3.3.7) and wall carvings (Figure 3.3.8) and therefore much less likely to have been created recently. This is without mentioning the Ica Stones of Peru, from the Incan civilization of the 13th Century (Figure 3.3.9). Comparing the skin patterns found from fossilized dinosaur skin indicate a spot-like pattern on some species, these look similar to the spots found on the Ica Stones (Figure 3.3.10).

Chapter 3: Ancient History Part II

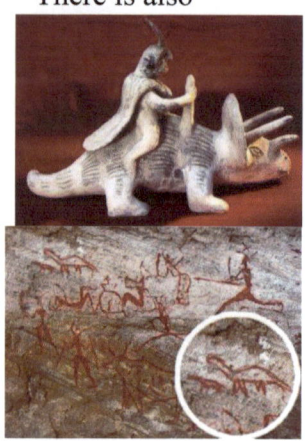

Fig. 3.3.10 Fossilized skin as compared to an Ica Stone

Fig. 3.3.11 Even more dinosaur art

There is also evidence that man interacted with these creatures because of the sheer number of artistic depictions found around the world (Figure 3.3.11). They were more than likely hunted to extinction.

The Carlisle Cathedral in England even has dinosaurs alongside so-called "extant species". These were encased around the tomb of Bishop Bell who died in 1496 (Figure 3.3.12).

The tomb not only includes sauropods, but also dogs, fish, birds, and eels. What reason would the artist have for showing one very particular species of an organism that had not even been discovered yet? Not only is this merely 500 years ago, but it is also almost 350 years before the word "dinosaur" was even coined.

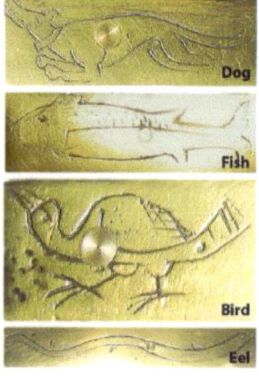

Since the dating method is questionable and there is evidence to suggest that there were dinosaurs that lived alongside man centuries ago, why would the textbook claim that they lived millions of years ago?

This chapter would be remiss if it did not mention anecdotal evidence to suggest that present-day travelers and natives in some lesser

Fig. 3.3.12 Sketching at the tomb of Bishop Bell

developed parts of the world have reported seeing creatures that could only be described as dinosaurs.

For example, a native in the jungles of Bolivia claimed to have seen an amazingly large armadillo (Figure 3.3.13A). However, when a westerner showed him a picture of an ankylosaur (Figure 3.3.13B), he claimed that was what he saw.

Fig. 3.3.13 Armadillo vs. Ankylosaur

Another example is the kongamato of Central Africa (Figure 3.2.14). According to the people of the region, it will eat any dead or decaying flesh that is left behind.

Fig. 3.3.14 Kongamato attacking

When asked to describe the creature, the natives will describe it as a large bat-like creature. When a westerner showed them a picture of a pterodactyl (Figure 3.3.15), it was immediately recognized as a kongamato.

Fig. 3.3.15 Pterodactyl

These communities have never heard of dinosaurs or that they are millions of years old.

Chapter 3: Ancient History Part II

Section 3
The First Cells:

The textbook will say that all life arose from inorganic materials in the ancient past. This, like many falsehoods in the textbook, has no evidence to support it.

Fig. 3.4.1 Living, Inorganic, and non-living

It is fitting to define the terms living, inorganic, and non-living. Living refers to any collection of matter that presently exhibits all of the signs of life (Figure 3.4.1A). Inorganic refers to any collection of molecules that do not exhibit the signs of life and was never alive (Figure 3.4.1B). Non-living refers to a collection of molecules that once exhibited the signs of life, but no longer does (Figure 3.4.1C).

Evolution postulates that over millions of years life arose from inorganic materials. If life did not spontaneously arise from inorganic materials, evolution over millions of years could never have occurred.

The textbook will say that the first life evolved out in the oceans millions of years ago.

Charles Darwin* proposed that the first life arose in a warm little pond. Thus, a single-celled organism evolved here which eventually became the ancestor of all organisms on Earth.

Unfortunately, Darwin did not understand hydrolysis (Figure 3.4.2). Hydrolysis is a chemical reaction in which water causes different compounds to break apart to form new smaller compounds. If the compounds combine to form a larger compound, it is known as condensation. Thus, hydrolysis is the reverse of condensation.

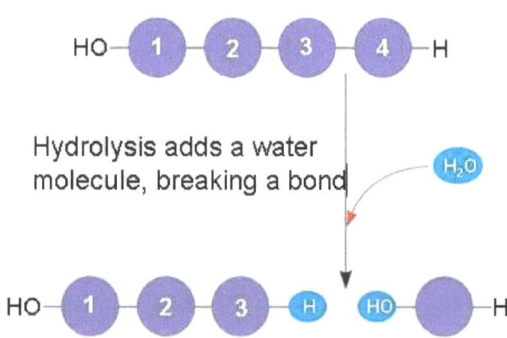

Fig. 3.4.2 Hydrolysis reaction

---

*Review Chapter 1 for more information on Darwin's qualifications and research. To remain consistent with current textbook teachings, we will juxtapose Darwin's research as a scientific analysis.

This is important because if there is water (a warm little pond) present, the condensation reaction does not occur. This is the equivalent of trying to reconstruct a salt crystal underwater.

The textbook will say that the Miller-Urey experiment of 1952 (Figure 3.4.3) provided the first piece of evidence that life arose from inorganic materials. However, this experiment was severely flawed.

The simple explanation of the experiment is that the scientists put some organic compounds (ammonia (NH_3), methane (CH_4), and hydrogen gas (H_2)) in a closed system and zapped evaporated water with electricity.

These compounds were not chosen randomly as Miller and Urey believed that these were the most common compounds in Earth's early history.

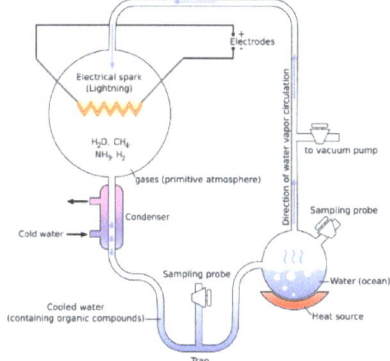

Fig. 3.4.3 The Miller Urey experiment

Let's take a closer look and examine why this experiment is fallacious. Firstly, it was conducted in a closed system, whereas the actual oceans are open systems.

Secondly, the amino acids had to be removed immediately otherwise they would have been destroyed by the sparks that supposedly would create even more life.

Fig. 3.4.4 A rusted chain

A much more glaring problem is that the experiment excluded oxygen since this would break down organic compounds, just like how iron rusts (Figure 3.4.4). Without oxygen in the atmosphere, Ultraviolet (UV) light breaks down ammonia. Since this was one of the compounds in their experiment, Miller and Urey could not have ammonia break down. Since the two excluded oxygen from their experiment, this apparatus does not accurately depict nature. This is without mentioning the hydrolysis that would break down the compounds in a "warm little pond".

There is no evidence to suggest that the Earth was ever free of oxygen. Even if the geologic column (from the previous chapter) was true, the lowest layers always have oxygen.

The results of the experiment are troubling as well.

The results produced 85% tar 13% carboxylic acid and 2% amino acids. The first two are toxic to life. If a compound is 98% toxic substance and 2% living cells, no creature will survive.

To reiterate, it is a fact that proteins do not come together in water (Figure 3.4.6). Therefore, the first amino acids could not have formed into proteins let alone into fully functioning cells.

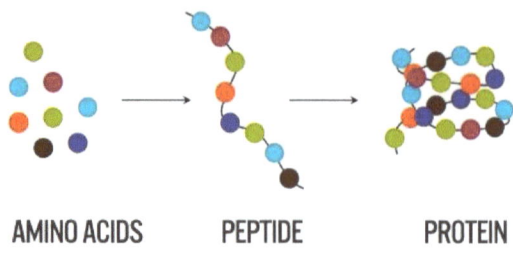

Fig. 3.4.6 A simple protein

If life cannot come about by natural forces over millions of years, the evolutionary timeline has no backing at all.

Not only do the molecules have to form against all odds, but they also have to form cells, which is impossible as well. (This will be further discussed in Chapter 6.) Even if the cells could form, they would still be single-celled organisms. These would have to become multi-cellular organisms over time.

The prevailing hypothesis is that the microbes began to collect together and take on specialized roles and work together, while this has been demonstrated in a lab, this is a far cry from a single organism (Figure 3.4.7).

The evolutionist will then go on to say that these organisms eventually collected to form colonies with much more complicated roles to form multi-cellular organisms. The sponge (Figure 3.4.8), is little more than specialized tissue although it is technically an animal. Because of this, series of assumptions, it is thought to be the first multi-cellular life form on earth.

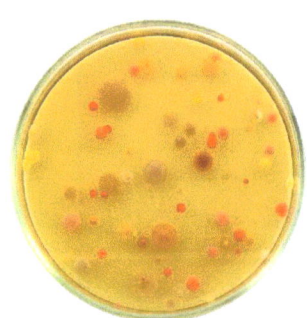

Fig. 3.4.7 A few colonies of bacteria.

There is still yet, an even bigger hurdle to pass. Every multi-cellular organism has a unique genetic code that is shared by every cell in its being. Colonies do not have this trait. Furthermore, the amount of DNA and cellular changes that occur between a single-celled organism and multi-cellular organisms are astronomical.

Fig. 3.4.8 A "simple" sponge.

Section 5
The Ages of Humanity:

The textbook will describe that humanity developed from simple ape-like creatures millennia ago. It will go on to describe that the development of tools began with simple stone tools that were chiseled into a shape to do a job (Figure 3.5.1) and eventually humans learned to make more advanced tools by metallurgy (Figure 3.5.2).

Fig. 3.5.1 A simple stone tool

Fig. 3.5.2 A complex metal tool

There is no proof that there was this development over time just that these tools were found. There are people even today that do not use automobiles, (Figure 3.5.3). If an anthropologist were to look at their civilizations now, they might just but still use a horse and buggy remains a century or two from assume that this group of people lived in the 19th Century.

Even today some people prefer to live out in the woods amongst nature and those that have lost skills and their way of life due to outside influences such as natural disasters and conquest. If the only person in the tribe who knows how to forge metal is killed, the rest of the tribe will not have this knowledge.

Fig. 3.5.3 A horse and buggy

Spanish conquest caused the Bororo, Pume and Guaja people of the Amazon Basin (Figure 3.5.4) to lose many

Fig. 3.5.4 The Amazon Basin

skills. However, they descended from people that made pottery. It is obvious to anyone studying the issue that skills can be lost and regained depending on the circumstances.

Chapter 3: Ancient History Part II

IN CONCLUSION:

There are a several ways to tell that the Earth is not millions of years old. (Section 1)
Many flawed methods of determining the age of the Earth have been used. (Section 2)
There are accounts of dinosaurs in artwork over the centuries. There are also anecdotal accounts in relatively unexplored parts of the world. (Section 3)
Life always comes from life and not from inorganic material. (Section 4)
There is no method by which single-celled organisms can become multi-cellular. (Section 4)
There is no proof that humanity went through different ages of development as the textbook describes. (Section 5)

Questions for Further Discussion:

1. Why would scientists limit themselves to only certain geologic clocks and be closed to those that imply a young Earth? (Section 1)
2. Are there any geologic clocks that point to millions of years? (Section 1)
3. Why would scientists use measurements that are known to be false? (Section 2)
4. Is there any scientific evidence that dinosaurs are millions of years old? (Section 3)
5. Why would dinosaurs be depicted in artwork alongside man around the world before the first fossil was discovered? (Section 3)
6. Why is it so important to the evolutionary hypothesis that life arose from inorganic material. (Section 4)
7. Why would scientists trust an illogical experiment to try to create life? (Section 4)
8. Why do scientists insist that colonies become unique organisms? (Section 4)
9. Why do scientists insist that there was a Stone Age?
10. Is there another hypothesis that will explain these phenomena?

Additional Reading

Religious Sources:

Dean, Don. "Did Man and Dinosaurs Co-Exist Thousands of Years Ago? If so, Where Is the Evidence?" *Biblical Truths*, 3 Feb. 2013, bibletruths2013.blogspot.com/2013/02/did-man-and-dinosaurs-co-exist.html.

"The Ica Stones." *The Greater Picture - Ascension / 2012*, thegreaterpicture.com/Ica_stones.html.

"The Ica Stones of Peru." *The Culture of the Hebrew Language*, www.ancient-hebrew.org/ancientman/1001.html.

"Modern Stone Age Reconsidered." Answers in Genesis, Answers in Genesis, 1 Sept. 1993, answersingenesis.org/human-evolution/cavemen/modern-stone-age-reconsidered/.

OFFICIAL, Kent Hovind. "9/24/18 -Dr. Kent Hovind: Interview with Joe Meyers - Dinosaurs Are Still Alive!" *YouTube*, Creation Science Evangelism, 24 Sept. 2018, www.youtube.com/watch?v=7HUr6eQtdcw&t=378s.

Parker, Chris. "OOPARTS(out of Place Artifacts)& ANCIENT HIGH TECHNOLOGY--Evidence of Noah's Flood?" *Top Eleven Mysterious Mysteries of the Pre-Columbian Americas That We Decided to Cram Into One Article....Page 54*, 2009, s8int.com/phile/dinolit87.html.

Rognstad, Matthew, and Timothy H Heaton. "The Age of the Earth and the Formation of the UniverseHonors Seminar (UHON 390), Fall 2005." *The Age of the Earth - Lord Kelvin's Heat Loss Model as a Failed Scientific Clock: Matthew Rognstad*, University of South Dakota, apps.usd.edu/esci/creation/age/content/creationism_and_young_earth/accelerated_decay.html.

Scharringhausen, Britt. "Is the Moon Moving Away from the Earth? When Was This Discovered? (Intermediate)." *Home - Curious About Astronomy? Ask an Astronomer*, Cornell University, curious.astro.cornell.edu/about-us/37-our-solar-system/the-moon/the-moon-and-the-earth/111-is-the-moon-moving-away-from-the-earth-when-was-this-discovered-intermediate.

"The 'Kongamato' of Africa." *Genesis Park Scriptural Evidence Comments*, Genesis Park, www.genesispark.com/exhibits/evidence/cryptozoological/pterosaurs/kongamato/.

Thomas, Brian. "Did Medieval Artists See Real Dinosaurs?" *The Institute for Creation Research*, 29 June 2018, www.icr.org/article/did-medieval-artists-see-real-dinosaurs

Wolpert, Stuart. "The Moon Is Older Than Scientists Thought, UCLA-Led Research Team Reports." *UCLA Newsroom*, University of California Los Angeles, 11 Jan. 2017, newsroom.ucla.edu/releases/the-moon-is-older-than-scientists-thought-ucla-led-research-team-reports.

Chapter 9: Starlight

NOTES:

NOTES:

CHAPTER 4
Radioactive Dating

Operation: Battleground Textbook

What Does the Textbook Teach?

    The textbook will say that if you measure the radioactive decay of certain elements that make up a substance, you can determine the age. (Section 1)

    The textbook will say that because of research in this field, the Earth is millions of years old. (Section 2 & Section 3)

Chapter 4: Radioactive Dating

Section 1
Why Dating Methods Fail:

One of the main problems with modern dating methods is that it is based on unverifiable assumptions.

Radioactive dating is a relatively simple procedure. The scientist measures the elements present in the compound. The element that decays is known as the Parent Compound and the element that is decayed is known as the Daughter Compound. Assuming that the reaction occurs at a constant rate, the scientist can hypothesize how old the sample is.

Carbon 14, for example, has a half-life of about 5,730 years and breaks down into Carbon 12.

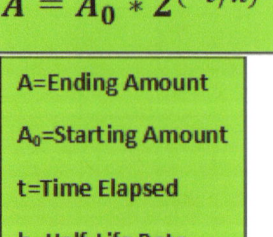

$$A = A_0 * 2^{(-t/h)}$$

A=Ending Amount

A_0=Starting Amount

t=Time Elapsed

h=Half-Life Rate

Fig 4.1.1 Half-Life formula

Fig 4.1.2 Water filling up a glass.

To accurately calculate the time elapsed, you need to know the starting amount (A_0), the ending amount (A), and the half-life rate (h) and assume that there is a closed system (Figure 4.1.1.)

Simply put, imagine a glass of water that is half full. Now imagine that water is dripping into the glass that fills it at an average of one-tenth of a glass per hour (Figure 4.1.2).

How long has water been dripping into the glass?

You might conclude that it has been on for five hours, but this is not necessarily true. You are assuming that there was no water in the glass to start with (starting amount), no water evaporated (closed system), and that the water has always been dripping at the same rate (Half-Life Rate).

A failure to account for any one of these assumptions can radically shift the time frame.

Let us compare this to radioactive dating.

Radioactive elements decay from one element (parent compound) to another element (daughter compound) (Figure 4.1.2). If there is any of the daughter compound present at the start of the reaction, it will look older than it is. This is the equivalent of having some water in the glass before the faucet was turned on. Potassium-Argon dating can be reset if more potassium is introduced to the system. Therefore if the system can be reset, there is no reason to believe that the reaction can be

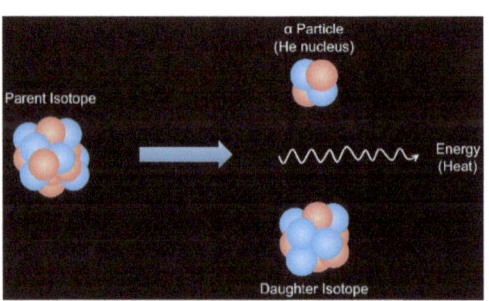

Fig 4.1.2 Radioactive decay illustrated.

measured empirically.

    We also have to assume a completely closed system. Any outside force that alters the environment can change the decay rate. If water washes away some of the compounds or a dog eats part of the sample, the numbers will change. In essence, we are assuming that no water evaporated or was poured out of our glass.

    Finally, we are assuming that the half-life rate was always the same and no factors were involved to change the rate. Circumstances that could change the rate are myriad including environmental pressures and weather. This would be the equivalent of changing the flow of the faucet during the experiment. It would illogical to conclude that all reactions will always occur at the same rate and pattern.

    Although there are a variety elements that hypothetically decay at different set rates that are used for diverse age ranges and for what the sample being tested is, Carbon-14 is the most well-known. The relationship between CO_2 and living things is best illustrated by the Carbon Cycle (Figure 4.1.3).

    Carbon-14 dating is based on the central assumption that the carbon intake into an organism by both consuming food and breathing out CO_2 should be roughly equal to the concentration of carbon in the atmosphere. This again is based on a number of assumptions. Such as, are there any circumstances that can disrupt the concentration of carbon in the atmosphere? Wind, man-made sources, volcanic eruptions, or a myriad of other sources can affect the concentration of carbon in the atmosphere over just the 5,730 years that the testing should not be reliable.

    Carbon-14 naturally breaks down into Carbon-12 at a hypothetical rate in relation to ultraviolet light amongst other factors. There is some research to suggest that the Earth would reach this equilibrium in 30,000 years. Therefore, if the Earth is millions of years old, the planet should have long reached equilibrium. The same research suggests that we are about 1/3 to equilibrium. This means that if we are to use Carbon-14 as our geologic clock (Chapter 3 Sections 1 and 2), the Earth is much closer to 10,000 years old.

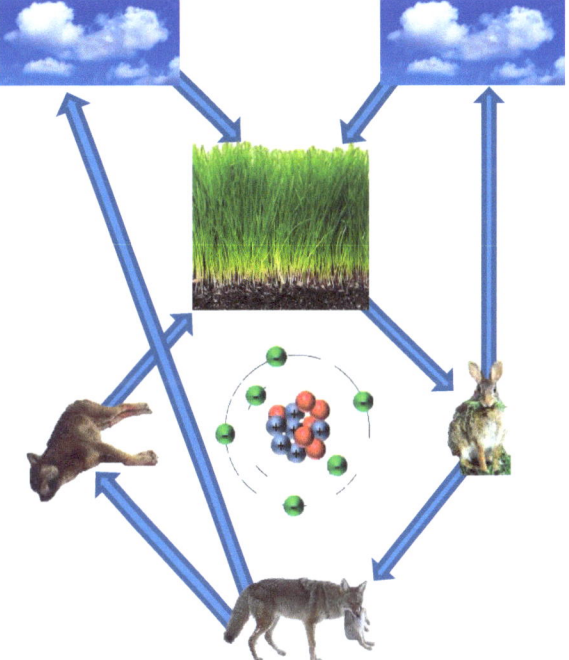

Fig 4.1.3 A simplified Carbon Cycle

Chapter 4: Radioactive Dating

There is some research to suggest that there are circumstances by which radioactive decay can be sped up. This would mean that a material that has a half-life of thousands of years could be reduced to tens of years. The applications in the disposal of radioactive waste are myriad (Figure 4.1.4).

Radioactive waste, if the radiometric dating method is to believed, can take years if not centuries to fully break down. If it is possible to shorten the time it takes for the sample to break down, nuclear power will be even cleaner and we can provide plenty of sustainable energy.

Since the decay rate is not static, it is not an accurate measure. Therefore, measuring radioactive decay is based entirely in unverifiable assumptions, the testing is unreliable.

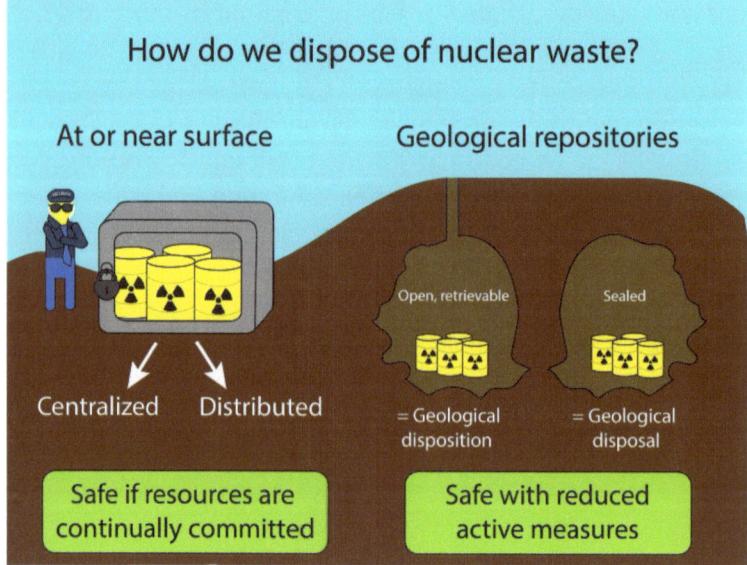

Fig 4.1.4 Nuclear waste disposal

Section 2
Where Dating Methods Fail:

Now that we have established that modern dating methods cannot be trusted from a theoretical standpoint, can they be trusted in practice?

There are a plethora of examples too numerous to list here. However, let us examine a few:

- The decay rate of Carbon 14 is about 5,730 years, but scientists have found statistically significant amounts in diamonds (Figure 4.2.1) that are supposedly millions of years old.

- The lower leg of the Fairbanks Creek mammoth (Figure 4.3.2) had an age of 15,380 years, while its skin and flesh were 21,300 years, according to Harold E. Anthony. This is a discrepancy of 72% or 5,920 years.

- Living mollusk shells were carbon dated as being 2,300 years old according to scientists M. Keith and G. Anderson.

- A freshly killed seal (Figure 4.2.3) was carbon dated as having died 1,300 years ago according to the *Antarctic Journal* in 1971.

- Shells from living snails were carbon dated as being 27,000 years old according to *Science*.

These are just a few examples. It seems that radioactive dating is not a reliable method since we cannot always observe the events as they occur. When events are unobserved, the dating method is never called into question.

If multiple dating methods give different ages, the sample is given an age that best fits the hypothetical geologic column. In essence, the fossils are used to date the rocks and the rocks are used to date the fossils.

Fig 4.2.1 A diamond containing C14

Fig 4.2.2 A mammoth dated at two different ages.

Fig 4.2.3 A living seal dated at thousands of years old.

Chapter 4: Radioactive Dating

Section 3
When To Use Dating Methods:

Based on what we have learned, it will come as no surprise to learn that most scientists do not use radioactive decay to date tissue, fossils, and rocks. This is because scientists realize how inaccurate radioactive dating can be. Despite this, it will still be taught in the textbook as if it were a scientific practice.

Radioactive dating is only used when the results support the evolutionary worldview. In general, if you were to take a sample to a lab to have the scientists date it, the scientists would first ask you what layer it was found in before dating it.

This is not to determine the environment, but it is so the scientist can find out which layer it was found in and thus to place it in the Geologic Column. (Review Chapter 2 for more about the inaccuracies regarding the Geologic Column) This is because the layer dates the fossils "more accurately" than any radioactive dating method.

Aside from sciences such as anthropology (the study of people and cultures), radioactive dating methods are rarely used. Even when anthropologists use radioactive dating, they will still only use the date that best fits the geologic column.

Different dating methods can provide very different results.

Fig 4.3.1 Mt. St. Helens

On May 18, 1980, Mt. St. Helens (Figure 4.3.1) erupted. Cooled lava was sampled. The lava from the dome was Argon dated to be .34-2.8 million years old. There is no field of science where such a large margin of error would be accepted as empirical, let alone scientific.

Polonium has a very short half-life (three minutes at most). The textbook will say that the Earth's surface was molten from its formation that eventually cooled down. However, there are polonium "halos" (Figure 4.3.2) that are present in stone such as granite. If the polonium was in stone that slowly cooled and hardened over millions of years, such halos should not exist.

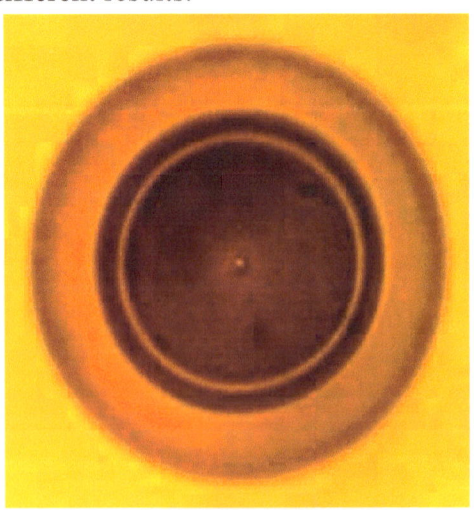

Fig 4.3.2
Polonium Halos

IN CONCLUSION:

In reality, radioactive dating methods are flawed based on unverifiable assumptions. (Section 1)

If dating methods cannot be supported by historical data, then the radioactive date is rejected. (Section 2)

Radioactive dating cannot be trusted since it can give dates that differ by hundreds or even thousands of years. (Section 3)

Questions for Further Discussion:

1. Why would scientists use a dating method that is unreliable as a standard? (Section 1)
2. Why would scientists use an unreliable dating method only when they do not have historical data to find the age? (Section 2 & Section 3)
3. Is there another hypothesis that will explain these phenomena?

Additional Reading

Secular Sources:

"Antarctic Journal." *Antarctic Journal*, vol. 6, Oct. 1971, p. 211.

Keith, M., and G. Anderson. "Science." *Science*, vol. 141, 1963, pp. 634–637.

R. E. Taylor and J. Southon, "Use of Natural Diamonds to Monitor 14C AMS Instrument Backgrounds," *Nuclear Instruments and Methods in Physics Research B* 259 (2007): 282–287.

Save-Soderberch, T., and I. U. Olsson. "C-14 Dating and Egyptian Chronology in Radiocarbon Variations and Absolute Chronology." Proceedings of the Twelfth Nobel Symposium, p. 35.

"Science." *Science*, vol. 224, pp. 58–61.

Sherriff, Lucy. "Astrophysicist Speeds up Radioactive Decay." *The Register® - Biting the Hand That Feeds IT*, The Register, 1 Aug. 2006, www.theregister.co.uk/2006/08/01/faster_decay/.

Additional Reading

Religious Sources:

Anthony, Harold E. "Nature's Deep Freeze." *Natural History*, Sept. 1949, p. 300.

Austin, Steve A. "Excess Argon within Mineral Concentrates from the New Dacite Lava Dome at Mount St. Helens Volcano." *The Institute for Creation Research*, 1996, www.icr.org/research/index/researchp_sa_r01/.

CMIcreationstation. "Radiohalos Ruin Radiometric Dating (Creation Magazine LIVE! 7-15)." *YouTube*, Creation Ministries International, 10 Oct. 2018, www.youtube.com/watch?v=xYPprMA4PrA.

D. B. DeYoung, *Thousands . . . Not Billions: Challenging an Icon of Evolution, Questioning the Age of the Earth* (Green Forest, Arkansas: Master Books, 2005), pp. 45–62.

Hovind, Eric. "Does Carbon Dating Prove the Earth Is Millions of Years Old?" *Creation Today*, Creation Today, creationtoday.org/carbon-dating/.

J. R. Baumgardner, "14C Evidence for a Recent Global Flood and a Young Earth," in *Radioisotopes and the Age of the Earth: Results of a Young-Earth Creationist Research Initiative*, eds. L. Vardiman, A. A. Snelling, and E. F. Chaffin (El Cajon, California: Institute for Creation Research, and Chino Valley, Arizona: Creation Research Society, 2005), pp. 587–630.

Snelling, Andrew A. "Carbon-14 in Fossils and Diamonds." *Answers in Genesis*, 1 Jan. 2011, answersingenesis.org/geology/carbon-14/carbon-14-in-fossils-and-diamonds/#fn_1.

Snelling, Andrew A. "Radiohalos." *Creation.com | Creation Ministries International*, Creation Ministries International, creation.com/radiohalosstartling-evidence-of-catastrophic-geologic-processes-on-a-young-earth.

Walker, Tas. "New Radiohalo Find Challenges Primordial Granite Claim." *Creation.com | Creation Ministries International*, Creation Ministries International, creation.com/new-radiohalo-find-challenges-primordial-granite-claim.

NOTES:

NOTES:

CHAPTER 5
Forming Elements

Chapter 5: Forming Elements

What Does the Textbook Teach?

The textbook will say that all the advanced elements in the universe were formed from fusion. (Section 1)

Evolution requires that the universe be millions of years old for there to be the time to make all of the larger elements. (Section 2)

Section 1
FORMATION OF ATOMS:

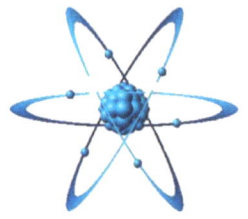

Fig. 5.1.1 A simplified model of the atom.

Atoms are phenomenally complex objects (Figure 5.1.1). You could spend a whole career studying the intricate and complex nature of atoms and never fully understand them.

For the millions of years of evolutionary history to make sense, scientists believe that the elements, which make up everything in the universe, need to have been formed in the hearts of stars.

Stars generate their vast power by fusing hydrogen atoms to create helium (Figure 5.1.2). This is a process known as nuclear fusion. Fusing two hydrogen atoms (one proton and one electron each) into a helium atom (two protons and two electrons) releases an astronomical amount of energy.

Now the bigger the atom, the more difficult it is to fuse. You can melt different types of metal, but they are still made from the same atoms that you started with. It does not make a new element. New elements require a more complicated process.

Going across the periodic table drastically increases the complexities of the atom (Figure 5.1.3).

Fusing atoms is a level of sophistication beyond all measure...

So far, scientists have been unable to find a star that can fuse past iron. Iron is only number 26 of 86 naturally occurring elements. The remaining 60 exist, but could not have formed by fusion, at least with the present model.

Fig. 5.1.2 Two hydrogen atoms fuse

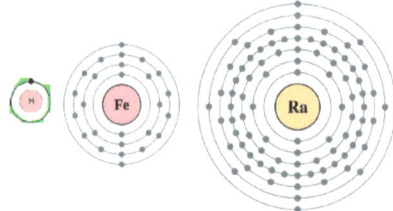

Fig. 5.1.3 Hydrogen, iron, and radium atoms

When the star starts to fuse to iron, the core will begin absorbing the energy faster than it can be put out. This results in stellar death.

A nova is an explosion that is generated when a star dies. A supernova is a larger explosion. Some scientists have suggested that multiple supernovae would create enough energy to create heavier atoms, but with all of the heavier elements that exist, such a claim is unfeasible.

If several stars have to be lost just to create the heavier elements that are relatively abundant, how are there any stars left?

In other words, the process of nuclear fusion is responsible for forming the elements above hydrogen. These elements then came together to form the stars. Stars are formed by the super-compression of elements. (This will be further discussed in Chapter 9.)

Chapter 5: Forming Elements

Fig. 5.1.4 The chicken and the egg.

Therefore, the elements are needed to make the stars, but the stars are needed to make the elements.

One cannot have come into existence by purely natural forces without the other. The elements would need to form so that they could produce a star and the star was needed to produce the elements (Figure 5.1.4).

Section 2
FORMATION OF THE UNIVERSE:

The universe, as we know it today is quite large and has levels of complexity that are difficult to even fathom.

The textbook will say that the Big Bang theory is the explanation for the formation of the universe. (Recall the definition of the word "theory" from Chapter 1.) To remain consistent, this chapter will reference this idea as the Big Bang Hypothesis.

This hypothesis posits that at some time in the distant past, the entire universe and all that it contains (all of the matter and all of the energy) existed as a singularity smaller than a period on this page (Figure 5.2.1).

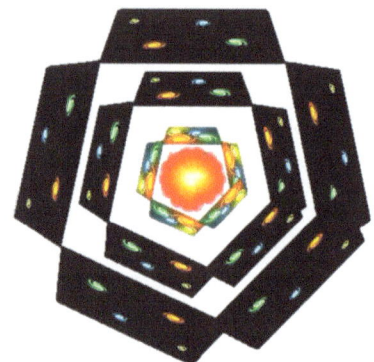

Fig. 5.2.1 The expansion of the universe

This singularity began to expand and eventually everything that we know today came into existence over slow and gradual processes. There is no scientific explanation for why the expansion of said singularity began or how said singularity came into existence.

If this hypothesis is true, the expansion occurred slowly and over, billions of years just like the fusion that created the heavier elements.

A glaring problem with this hypothesis is the force of gravity. Anyone who has watched a space shuttle launch can tell how difficult it is to break through the Earth's atmosphere. This is comparable to expanding gasses (Figure 5.2.2).

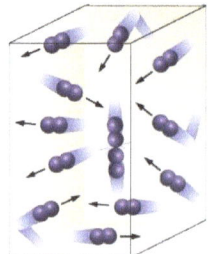

Fig. 5.2.2 Gasses expand

Put another way, gravity naturally pulls all matter together. Considering the vast gravity that exists even on small planets such as Earth, how much pull would all the matter in the universe have?

Imagine all the matter in the universe in a single dot and how much gravity this would hold. If the laws of physics known presently hold true, the expansion should not have occurred.

Chapter 5: Forming Elements

IN CONCLUSION:

Even just the atom is amazingly complex and forming elements requires tremendous amounts of energy. There is no scientific or natural method by which elements past iron can be formed. (Section 1)

Since the stars are needed to form the elements and the elements are needed to form the stars, one could never have existed without the other. (Section 1)

Saying that the universe formed by a rapid expansion does not coincide with any experimental science and is not backed by empirical research. (Section 2)

Questions for Further Discussion:

1. If we see iron and many other heavier elements across the cosmos, how did these elements come into existence? (Section 1)
2. Which came first the elements or the stars? (Section 1)
3. How did the universe come into being if not from a rapid expansion? (Section 2)
4. Is there another hypothesis that will explain these phenomena?

Additional Reading
Chapter 5 Forming Elements

Secular Sources:

Pandian, Jagadheep D. "How Are Light and Heavy Elements Formed? (Advanced)." *Home - Curious About Astronomy? Ask an Astronomer*, Cornell University, curious.astro.cornell.edu/about-us/84-the-universe/stars-and-star-clusters/nuclear-burning/402-how-are-light-and-heavy-elements-formed-advanced.

Religious Sources:

Chapter 5: Forming Elements

NOTES:

CHAPTER 6
Genetic Changes

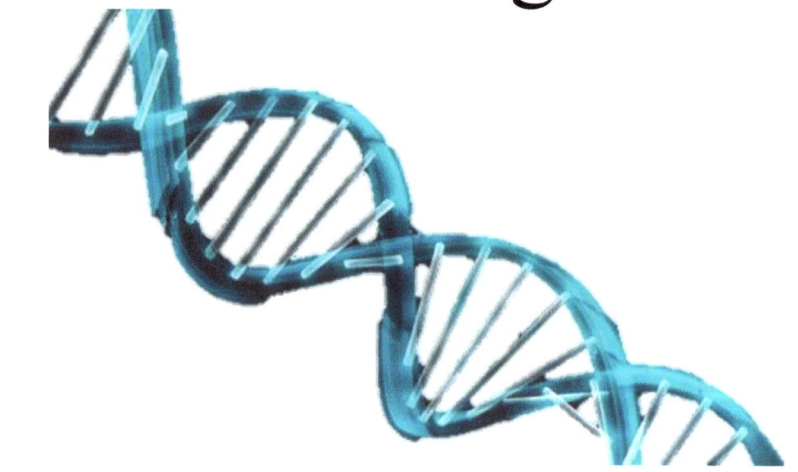

What Does the Textbook Teach?

The textbook will teach that small changes over long periods of time is what it takes to change from one type of organism to another. The textbook will define this as evolution. (Section 1, Section 2, and Section 4)

It will teach you that two major pieces of evidence for evolution are microbes gaining immunity to disease and mutations in humans. (Section 3 & Section 5)

Chapter 6: Genetic Changes

Section 1
DARWIN'S FINCHES:

When Charles Darwin* went to the Galapagos Islands, he found several species of finch (Figure 6.1.1). He noticed that each of their beaks were slightly different depending on their diet.

Because of this, Darwin* concluded that humans and pine trees shared a common ancestor. In his book, *Origin of Species*, he

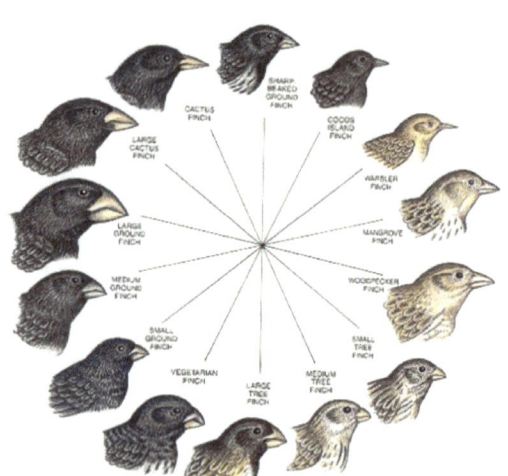

Fig. 6.1.1 Variations of finches

describes the similarities amongst different birds as proof that all living organisms on the face of the Earth share a single common ancestor (Figure 6.1.2).

Darwin declared:
"It is a truly wonderful fact that all animals and all plants throughout all time and space should be related to each other."

Recall from Chapter 3 that Darwin* also insisted that all life evolved from non-living materials.

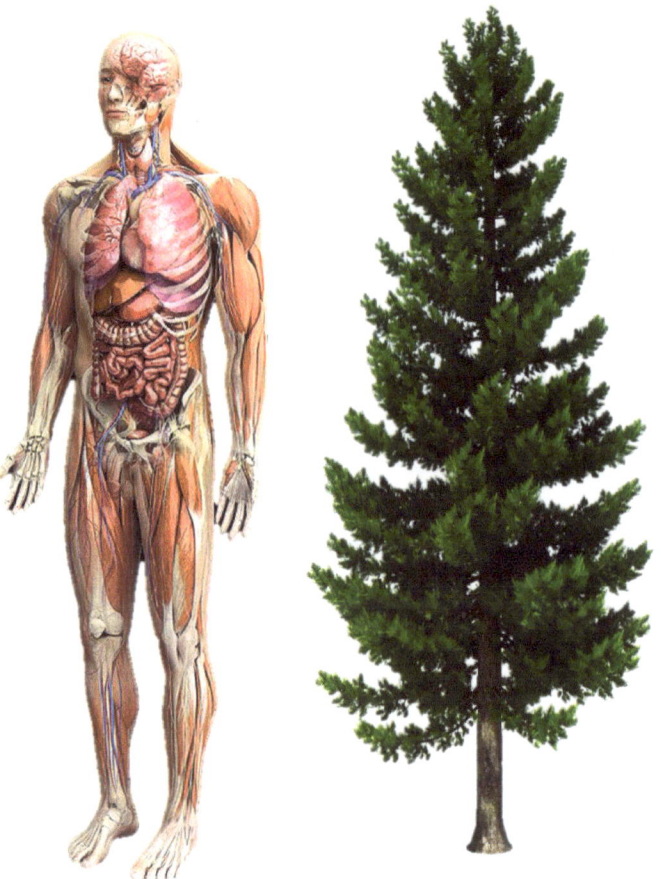

Fig. 6.1.2 A human and a pine tree

*Review Chapter 1 for more information on Darwin's qualifications and research. To remain consistent with current textbook teachings, we will juxtapose Darwin's research as a scientific analysis.

Section 6.2
GENETIC VARIABILITY:

Scientists once believed that 1% of our DNA was different from apes (Figure 6.2.1 AB). However, scientists now believe the difference may be as much as 5% (Figure 6.2.1 C). Another shocking find was that the researchers excluded 18% of the ape DNA and 25% of the human DNA when making the comparison. This would mean that we share roughly 56% of our DNA with apes. This means that there is a 44% genetic variation between humans and apes (Figure 6.2.1 D).

Mice, dogs, and chickens have also had their genomes compared to humans (Figure 6.2.2). All three creatures possess more genetic similarities to humans than humans do to apes. Yet no evolutionist that will say that humans evolved from mice.

The original Mona

Mona Lisa with 1% variation.

Mona Lisa with 5% variation.

Mona Lisa with 44% variation.

Fig. 6.2.1 The Mona Lisa with slight variations.

Chapter 6: Genetic Changes

Consider how genetic similarities are an insufficient standard to use. Furthermore, these similarities depend entirely on which traits that are being compared.

Imagine two organisms that each have four genes each (Figure 6.2.3). If you were to compare the genes from 1D and 2D, you would see these organisms as 100% similar. However, if you compared a larger section of the genome such as 1C and 1D and then compared it to 2C and 2D, you would conclude that the organisms are 50% similar. The fact of the matter is, that depending on which set of genes you look at, changes how similar the organisms are.

Fig. 6.2.2 Genetic similarities between a mouse (90%), dog (84%), and chicken (65%) as compared to humans

Consider that each of the billions of genes do not simply code for a single trait, but many traits that interact in different ways with each other. (epigenetics).

Despite this setback, many scientists prefer to look at chromosomes.

To understand chromosomes, know that any trait for any species has the trait on an analogous chromosome. This chromosome count is species specific. For example, human blood type is contained on Chromosome 9. This is true for all humans, past and present.

Although we do share many traits with apes and other simians, we also possess a different number of chromosomes.

Humans have 23 pairs of chromosomes and apes have 24 pairs. That may not seem like much until you realize exactly how much information is stored in just one chromosome. A human chromosome contains about 1,400 genes. This means that there are at least 1,400 gene pairs that just appeared and switched themselves across different chromosomes between different species of simians and humans.

To further complicate things, the analogous traits of humans are rarely found on the same chromosome as they are in other simians.

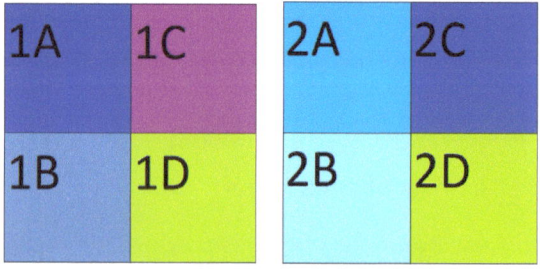

Fig. 6.2.3 Which traits are being compared?

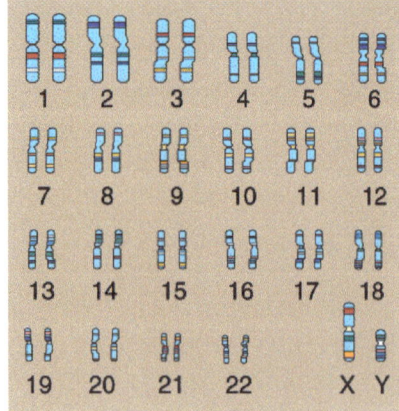

Fig. 6.2.4 The 23 pairs of human chromosomes.

If we were to compare the DNA code to the text of a book, each pair of chromosomes would be a different chapter. Thus, we would have 23 chapters (Figure 6.2.4) and the apes would have 24 chapters.

In this analogy, the genetic code that makes up the organism would be the sentences found in each chapter.

We should look for similar sentences in both the human book and the ape book. However, the sentences that are found together in the human book are found in different chapters in the ape book.

This would mean that over 1,400 pairs of genes (sentences) would have had to switch across many chromosomes (chapters) as the transition was made from apes to humans. Even if these 1,400 gene pairs did switch over many generations, there is no empirical data or any other evidence to show these transitional forms. Transitional forms are hypothetical organisms that express physical traits of both an ancestral species and those of the one that they are evolving into (This will be further discussed in Chapter 7).

This large genetic variance is a monumental change, to say the least. We see the genes are not on analogous chromosomes with a number of other simians and humans, such as gorillas (Figure 6.2.4). Scientists have mapped the chromosomes of multiple simians and humans and we see that genes are tied to specific locations.

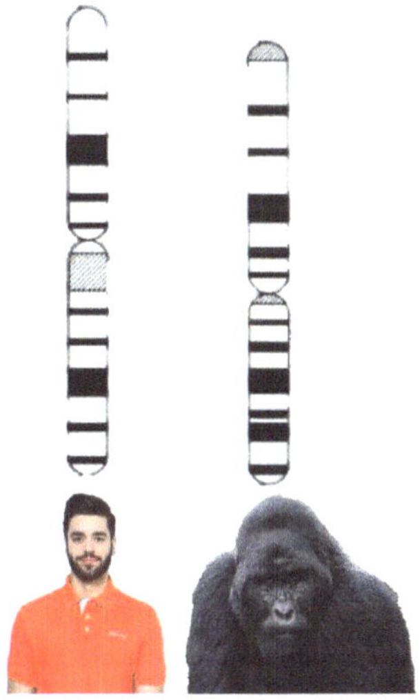

Even though that the chromosomes show no commonalities on the scale that evolution would require, many scientists will still falsely teach that the DNA between apes and humans 99.9% similar.

We see the same gene configuration when we look at virtually any other species. For example, fruit flies have eight chromosomes and house flies have twelve. If we assume an average of 10,000 genes per fly chromosome, the house fly has about 40,000 more genes than the fruit fly. That large number of genes and switching across chromosomes would be fatal to the fly or any organism.

Fig. 6.2.4 Chromosome 1 in Humans and Gorillas

For example, there are 40,000,000 (40 million) mutations between chimpanzees and humans. A mutation is any change in the DNA code. Even with a generous 100 mutations per generation, these changes would need 400,000 generations to complete (Figure 6.2.5). If a new generation was spawned every ten years on average, this would mean that the time between the two species would still be 4 million years. Realistically, the changes between each generation is closer to 42 genes across all traits. This is 100 traits amongst the differences between apes and humans. Therefore, even under the most favorable conditions, there is not nearly enough time to fit within the evolutionary hypothesis to account for all of the variations between apes and humans. Thus, there is a staggering absence of these countless transitional forms (This will be further discussed in Chapter 7.).

Generation	Year	Remaining Genes	
		Number of Ape Genes Remaining	Percent of Ape Genes Remaining
1st	202,000 BC	40,000,000	100.00000%
2nd	201,990 BC	39,999,990	99.99998%
3rd	201,980 BC	39,999,980	99.99997%
4th	201,970 BC	39,999,970	99.99997%
5th	201,960 BC	39,999,960	99.99997%
399,995th	3,797,950 AD	500	12.50000%
399,996th	3,797,960 AD	400	10.00000%
399,997th	3,797,970 AD	300	7.50000%
399,998th	3,797,980 AD	200	5.00000%
399,999th	3,797,990 AD	100	2.50000%
400,000th	3,798,000 AD	0	0.00000%

Fig. 6.2.5 The mutations from ape to human

Now it was mentioned earlier in the chapter that not only do apes and other simians have a different chromosome count than humans, but the genes are not on analogous chromosomes. Now the evolutionist will attempt to explain that humans have one less pair of chromosomes by saying that our chromosome two is a fusion of two different chromosomes as if the evolutionist observed this event happening.

Chapter 6: Genetic Changes

First it is important to explain some of the more intricate details of a chromosome (Figure 6.2.6). A Chromosome is broadly made of three main parts: two telomeres, two coding portions, and one centromere. The telomere (these will be further discussed in Chapter 8) provide non-coding DNA as a buffer during cell division and are found at both ends of the chromosome. The centromere acts as a union point between a chromosome pair. Intermixed between these two parts is all of the coding portion of the DNA.

Now the textbook will say that human Chromosome 2 has an active and an inactive centromere with four sets of telomeres (Figure 6.2.7). However, the telomere is millions upon millions of sequences in the two sets of "telomeres" in the Chromosome 2 are only a few sequences in length.

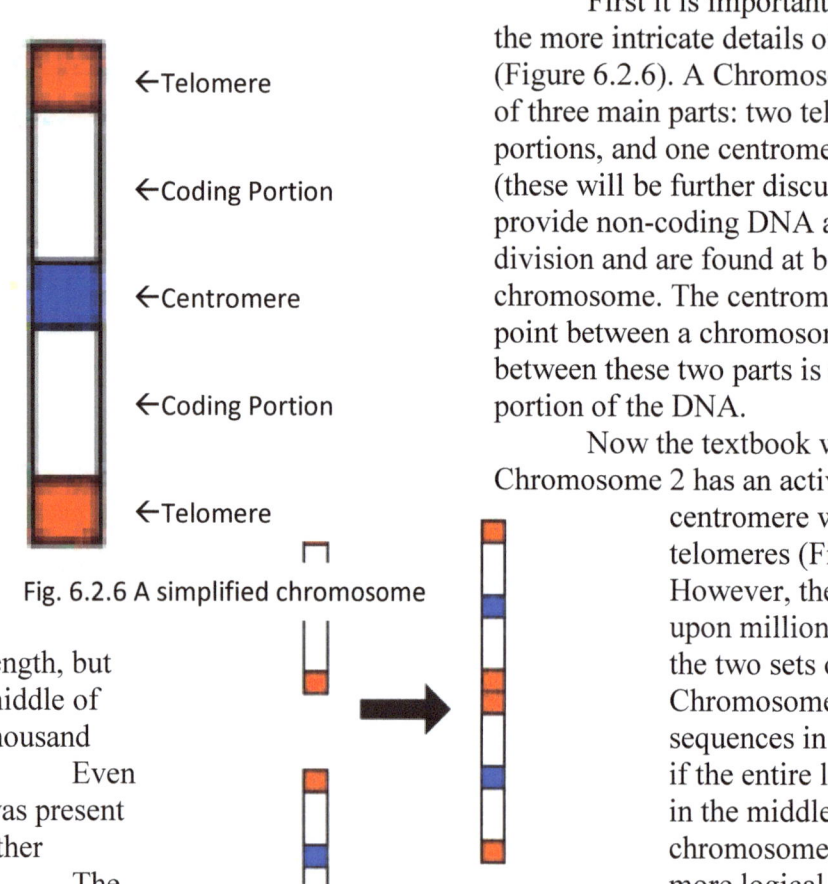

Fig. 6.2.6 A simplified chromosome

length, but middle of thousand
    Even was present other
    The humans and ancestor.

Fig. 6.2.7 The supposed fusion of the chromosomes

if the entire length of the telomere in the middle of the chromosome, chromosomes share this trait. more logical answer is that apes do not share a common

-78-

To further entangle this genetic web, even DNA is not the only change that was needed, the protein expression is also a concern. To make a very long explanation short, genes not only code for proteins but can also affect how other genes are expressed. This layer of complexity is known as epigenetics.

For example, both caterpillars and butterflies are the same species, but one becomes the other. The differences are just a matter of how the traits are expressed (Figure 6.2.8).

Fig. 6.2.8 The butterfly metamorphosis

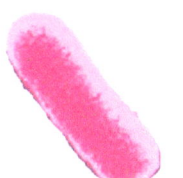

Fig. 6.2.10 An E. Coli

Even the smallest of bacteria are amazingly complex. Halophiles (salt-resistant) and thermophiles (heat-resistant) bacteria that are supposedly the closest to the ancestral archaebacteria are incredibly complex and show a many very advanced cell structures (Figure 6.2.9). E. Coli has 4,639,221 nucleotide base pairs which corresponds to 4,288 genes (Figure 6.2.10). Mycobacterium genitalium is known as one of the simplest organisms on Earth. Even so, it still has 580,000 base pairs and 482 genes.

Fig. 6.2.9 A hot spring where thermophile bacteria live.

Fig. 6.2.11 Richard Dawkins

To prove the possibility of genetic formation by natural processes, Richard Dawkins (Figure 6.2.11) once posited that if a monkey had a keyboard with 27 keys (26 letters and a space bar) and randomly typed letters, he would eventually type a complex phrase such as "ME THINKS IT BE A WEASEL". The monkey has a 1/27 or 3.7% chance of hitting any one key. In reality, the odds of typing just this one simple sentence in a row without mistakes is 1/2,252,839,954,493,917,441,184,014,787,477,300 (1/2.25 decillion) or .000000000000000000000000000000444%.

Dr. Dawkins gets around these astronomical odds by suggesting that if the correct key was pressed, it would remain locked and not need to be retyped.

This brings about two major problems:
1. It implies that the result is known and planned for. Simply put, setting the keys implies that the first gene sequence that was needed for survival was put in place and just waited for the remaining sequences.
2. It implies that the sequence and the cellular component that it coded for would just wait for all of the other vital components. In other words, the heart would just sit around and wait for the veins and arteries to form. Without the other components, the individual parts would not survive.

Chapter 6: Genetic Changes

Section 3
ANTIBIOTIC RESISTANCE:

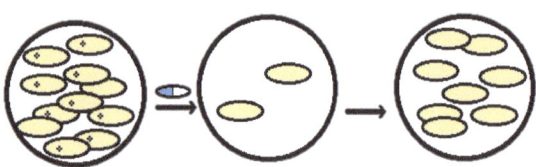

Fig. 6.3.1 Bacteria "gaining" immunity

The textbook will say that one of the best pieces of evidence for evolution is immunity to drugs in bacterial populations. If you are given an antibiotic and do not use the drug until all the bacteria are dead, those that are still alive will gain immunity to the drug (Figure 6.3.1).

To become immune to anything, a bacteria must lose the portion of the genetic code that the antibiotic effects. This means that it loses some part of what makes it work (Figure 6.3.2).

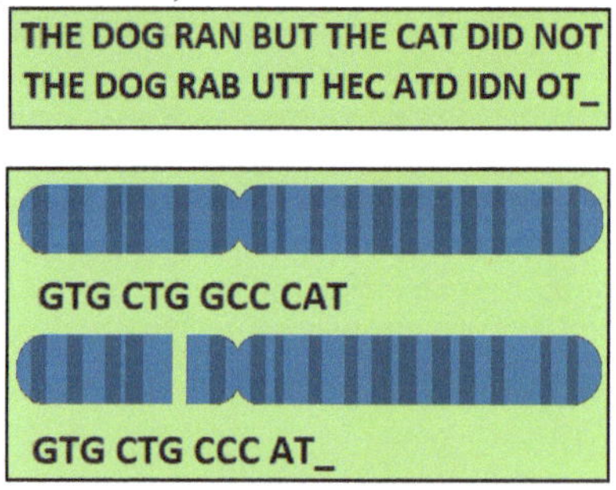

The part that the bacteria is losing is sometimes how effective it is at absorbing nutrients from its surroundings. In this way, the immune bacteria are much better at surviving in a sterile environment than their counterparts. However, if the same bacteria were in any other environment, it would not be able to take in nutrients as efficiently as its "vulnerable" counterparts. Therefore, the bacteria that are not immune would multiply faster.

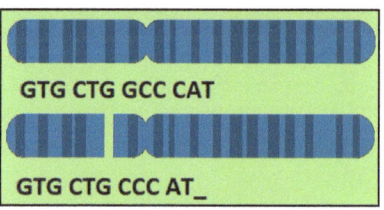

Fig. 6.3.2 Deleted segments of a gene.

Therefore, the bacteria are not gaining any information, but are losing information.

A simple analogy is that losing teeth ensures that someone will not have cavities, but the same person will then be unable to

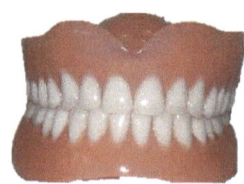

Fig. 6.3.3 False Teeth

eat certain foods (Fig. 6.3.3). Just as the bacteria lost information that makes weaker than its counterparts that are not immune, information has been lost.

Losing information is the very opposite of what is needed for evolution because this requires an increase in genetic information, which is not observed. Evolution requires new DNA, but with microbial immunity, all we see is a decrease in genetic diversity.

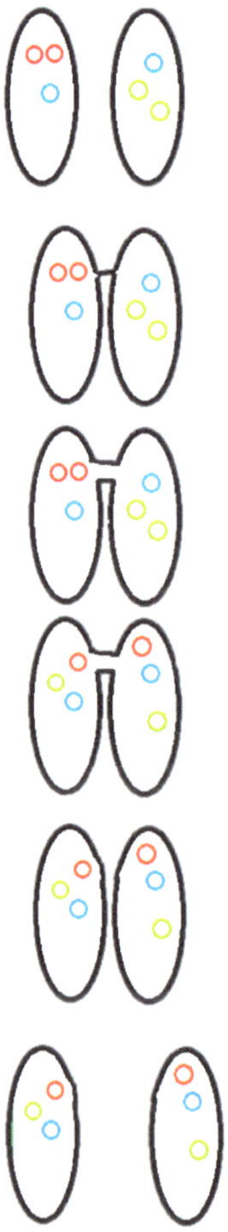

Another method of developing resistance is gene shuffling between bacteria (Figure 6.3.4). Bacteria can trade genes between one another. However, this is still just a variation within what was already present and no new genetic material was created.

Fig. 6.3.4 Bacterial gene transference

Section 4
VARIATIONS AND HYBRIDS:

Fig. 6.4.1 A Polar bear and a Grizzly bear.

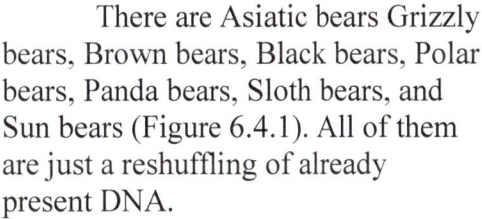

Fig. 6.4.2 A Grolar Bear

There are Asiatic bears Grizzly bears, Brown bears, Black bears, Polar bears, Panda bears, Sloth bears, and Sun bears (Figure 6.4.1). All of them are just a reshuffling of already present DNA.

Pizzly bears and Grolar bears are the offspring of Grizzly bears and Polar bears (Figure 6.4.2). That is because both are variations of a bear.

There are all sorts of dogs in the world. We have big dogs and little dogs, but they are all dogs (Figure 6.4.3). There are animals as big as elephants and as small as fleas, but there are still limits as to how far a species can diversify.

Fig. 6.4.3 A Great Dane and a Chihuahua

Wolves, dingoes, and Huskies (Figure 6.4.4) are all likely descended from a common ancestor. One of the main differences between the three is hair length.

If true, wolves (medium length hair) moved to climates that were either frigid or warm. A variation to Huskies (long hair) dominated colder areas and dingoes

Fig. 6.4.4 A dingo, wolf, and Husky

(short hair) came to live in warmer climates. Variations such as these occur in all species due to environmental pressures, but these variations have limits.

All that has been shown is a variation of already existing genetic material. The textbook will say that organisms have drastically changed over the years, but there is no fossil record or any indication to confirm this. (This will be further discussed in Chapter 7.)

Chapter 6: Genetic Changes

Section 5
VARIATIONS IN HUMANS:

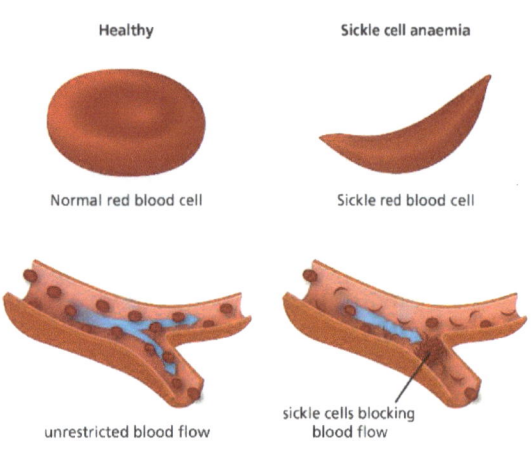

Fig. 6.5.1 Sickle Cell Anemia and its effects.

Scientists will posit that one of the most common examples of evolution in humans is Sickle Cell Anemia (Figure 6.5.1). This mutation is caused by a single letter of variation in the DNA. This causes the red blood cells to take on a new appearance that keeps them from carrying oxygen as efficiently (Figure 6.5.2). The textbook will say that this is an adaption to help those in mosquito-infested areas of Africa.

This is far from evolution since this is nothing more than a variation of already existing DNA.

To take this train of thought to its only logical conclusion: If you cut off your foot, you would never get athlete's foot. Therefore, a person born without feet would have an evolutionary advantage.

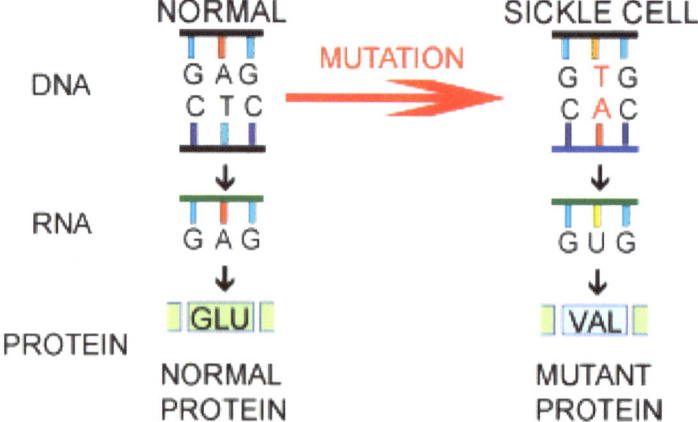

Fig. 6.5.2 The mutation of Sickle Cell Anemia.

IN CONCLUSION:

Darwin concluded that humans and pine trees were related because there were birds with slightly different beaks. (Section 1)

We do not see new information appearing on the DNA code. If we compare where the information is on the analogous chromosomes, it is illogical to assume common ancestry. (Section 2)

There are limits to how far genetic variation can go. These limits keep all creatures from having a single common ancestor. (Section 3 & Section 4)

All that has ever been observed is a variation of already existing genetic material. Sometimes the genes are varied or deleted, but this is not the formation of new genes. (Section 4 & Section 5)

There is no evidence of a fused chromosome.(Section 6)

Questions for Further Discussion:

1. Why would Darwin conclude that humans and pine trees are related? (Section 1)
2. Why would scientists say that species are directly related to one another when they have different chromosome counts and epigenetics show a much larger variation in genetic diversity? (Section 2)
3. Why is it so important that the chromosomes fuse during the transition from apes to humans? (Section 2)
4. Why would scientists say that these mutations (such Sickle Cell Anemia and antibiotic resistance) are evidence for evolution when it is simply an example of deleted or reshuffled information? (Section 3 & Section 5)
5. Is it truly evolution if you are just losing that which makes you vulnerable to disease? (Section 3 & Section 4)
6. If just one amino acid in the code for a red blood cell can create Sickle Cell Anemia, how much more chaos would be caused by changing thousands of genes simultaneously? (Section 5)
7. Is there another hypothesis that will explain these phenomena?

Additional Reading
Chapter 6 Genetic Changes

Secular Sources:

"Fruit Fly Genetics: Chromosomes & Genes." *Orkin.com*, www.orkin.com/flies/fruit-fly/fruit-fly-genetics/.

Garrett-Hatfield, Lori. "Animals That Share Human DNA Sequences." *Education*, Seattle PI, 29 Sept. 2016, education.seattlepi.com/animals-share-human-dna-sequences-6693.html.

"House Fly Chromosome Number." *Via Lima Chicago*, www.vialimachicago.com/house-fly-chromosome-number.

MinuteEarth. "Are We Really 99% Chimp?" *YouTube*, YouTube, 11 June 2015, www.youtube.com/watch?v=IbY122CSC5w.

Darwin, Charles. *The Origin of Species by Means of Natural Selection: or the Preservation of Favored Races in the Struggle for Life*. D. Appleton and Co., 2006.

National Center for Biotechnology Information (US). "Chromosome Map." *Current Neurology and Neuroscience Reports.*, U.S. National Library of Medicine, 1 Jan. 1998, www.ncbi.nlm.nih.gov/books/NBK22266/.

Perry, Jon. "Human Mutation Rate: How Many DNA Mutations Happen Each Generation?" *Stated Clearly*, Stated Clearly, statedclearly.com/articles/human-mutation-rate-how-many-dna-mutations-happen-each-generation/.

Religious Sources:

Anderson, Daniel. "Decoding the Dogma of DNA Similarity." *Creation.com | Creation Ministries International*, Creation Ministries International, 6 June 2007, creation.com/decoding-the-dogma-of-dna-similarity.

Bergman, Jerry. "Does the Acquisition of Antibiotic and Pesticide Resistance Provide Evidence for Evolution?" *Creation.com | Creation Ministries International*, Creation Ministries International, creation.com/does-the-acquisition-of-antibiotic-and-pesticide-resistance-provide-evidence-for-evolution.

Purdom, Georgia. "Antibiotic Resistance." *Answers in Genesis*, answersingenesis.org/natural-selection/antibiotic-resistance/.

Purdom, Georgia. "Are Humans and Chimps Related? - Creation Museum Live! | September 5, 2019." *YouTube*, Answers in Genesis, 5 Sept. 2019, https://www.youtube.com/watch?v=j24zXE_G3_w.

NOTES:

CHAPTER 7
Transitional Forms

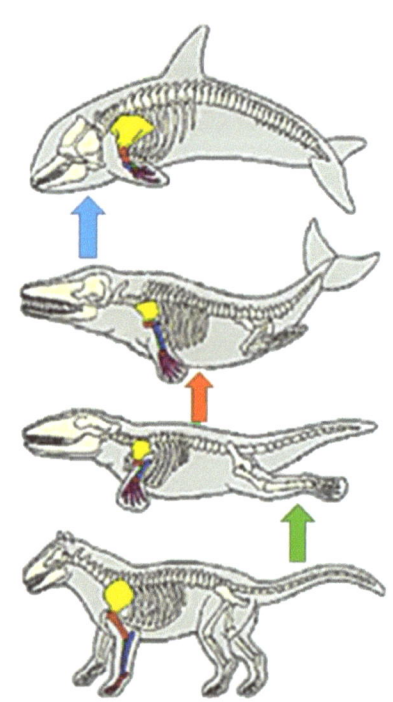

Chapter 7: Transitional Forms

What Does the Textbook Teach?

The textbook is going to teach that all organisms descended from a common ancestor. The only way that this could occur is with millions of years. (Section 1)

The textbook describes how small changes and variations that have occurred over the years are evidence for these transitional forms. (Section 2)

Some evolutionists believe that dinosaurs are the transitional form between reptiles and birds (Section 3).

The textbook will say that small incremental changes throughout human history shows how we changed from ape-like ancestors. (Section 4)

The textbook will say that we can chart the changes in species over the millennia. (Section 5)

Scientists claim that because the shape and arrangement of bones found in many vertebrates are similar, the vertebrates share a common ancestor. (Section 6)

Section 1
Transitional Forms:

Fig. 7.1.1 The supposed changes in humans.

The textbook will teach that all organisms evolved from a single common ancestor millions of years ago (Figure 7.1.1).

Since we supposedly share this common ancestor, the textbook will say that there are many transitional forms. A transitional form is a step along the evolutionary chain in which one will exhibit traits of both organisms.

Therefore, some form or variety of each organism is closest to its ancestral species. For example, the Alaskan Malamute (Figure 7.1.2) is among one of the closest to the wolf, from which all domesticated dogs descended.

Fig. 7.1.2 An Alaskan Malamute

Taking this line of reasoning to its only logical conclusion, some varieties of humans are closer to our ape-like ancestors than others. This has been the philosophy of many dictators (Figure 7.1.3) and eugenicists (Figure 7.1.4) around the world. A eugenicist is someone who believes that one group of people is genetically superior to another and that their DNA corrupts the rest of the gene pool.

Fig. 7.1.3 Adolf Hitler

Perhaps the most infamous was the case of Ota Benga (Figure 7.1.4). In 1906, Ota Benga was a pygmy that was placed in an exhibit at the Bronx Zoo to show how far humanity had evolved from ape-like ancestors. Specifically, to show how similar the people of Africa were to simians.

Going back to the diaries of slave owners from centuries ago, we see that slaves were always treated as less than human. This racially charged barbarism was only made worse when

Fig. 7.1.4 Margaret Sanger

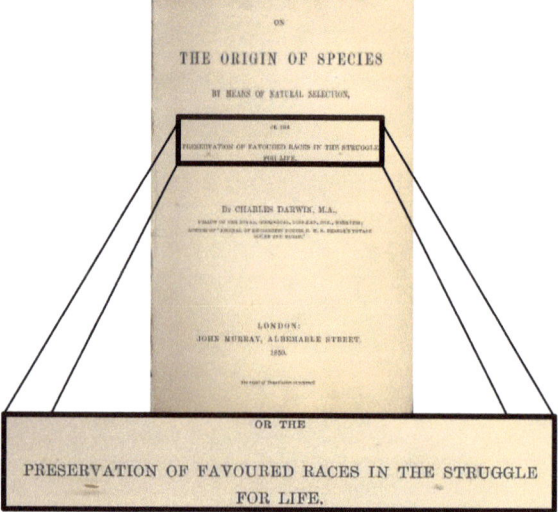

Fig. 7.1.5 Darwin's original book.

fueled by the addition of millions of years and Darwinian evolution (Figure 7.1.5).

Fig. 7.1.4 Ota Benga

Another moral atrocity was committed in Australia. European scientists found aboriginal graves and began robbing them to put the skulls in museums to prove the evolutionary hypothesis true (Figure 7.1.6).

There are other atrocities as well, but these few will suffice to demonstrate how far Darwinian supporters went to back up their hypothesis of transitional forms.

Fig. 7.1.6 A modern Australian Aborigine

Section 2
The History of Animals:

The textbook will say that aquatic mammals such as whales and dolphins descended from a land-dwelling ancestor. Their evidence consists only how these creatures swim. Since most other aquatic creatures propel themselves forward by moving their tails from side to side (Figure 7.2.1A), they must have a different ancestor than one that moves its tail up and down (Figure 7.2.1B). This means that the animals evolved on to land from fish that somehow grew legs and then ventured back into the water.

If this is true, we would expect to see some fossil evidence of this change from a land-dwelling creature to a sea-dwelling creature. In reality, no such creature has been found.

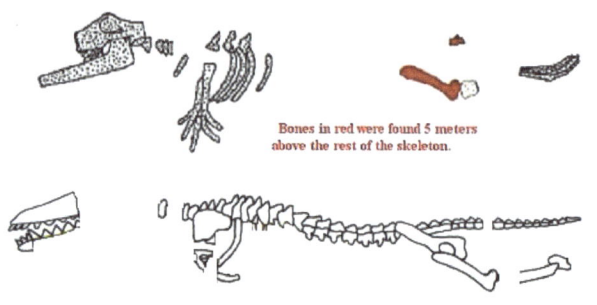

Fig. 7.2.2 The "pakicetus" bones

Fig. 7.2.1 How aquatic creatures swim.

One of the fossils that the textbook will mention is the infamous, pakicetus. This is a strange find to be used by the evolutionists since scarcely any bones were found (Figure 7.2.2A). With so many bones falsified by researchers to fill in the gaps (Figure 7.2.2B) in the fossil, it is hardly scientific to call this a transitional form.

This hypothetical creature was formed from less than half of a fossil and then labeled a transitional form without evidence. For example, the pelvis, and the majority of the vertebral column are missing. These two are vital parts in determining how a creature moved.

Fig. 7.2.3 A pakicetus recreation

The textbook will say that this transitional form is undeniable proof of evolution despite the drawings of the creature having almost no fossil evidence as source material (Figure 7.2.3).

Another point of contention that comes about when going from a land-dwelling mammal to a sea-dwelling mammal is the mouth.

In land-dwelling mammals, the mouth separates into two pathways. One leads from the mouth to the lungs for respiration and the other path from the mouth to the stomach for digestion.

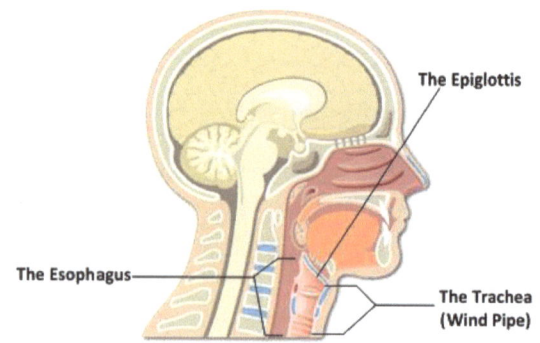

Fig. 7.2.4 The Epiglottis

Chapter 7: Transitional Forms

There is a valve known as the epiglottis (Figure 7.2.4) that prevents food from going into the lungs.

Sea-dwelling mammals have the mouth break into a single pathway. (Figure 7.2.5). This leads from the mout to the stomach for digestion. There is a separate path from the blowhole to the lungs.

It should come as no surprise that both of these had to come into existence simultaneously for the creature to survive underwater.

There is no fossil evidence to show these two separating into two distinct passages.

Furthermore, without these two separate pathways becoming fully functional concurrently, the creature would suffocate.

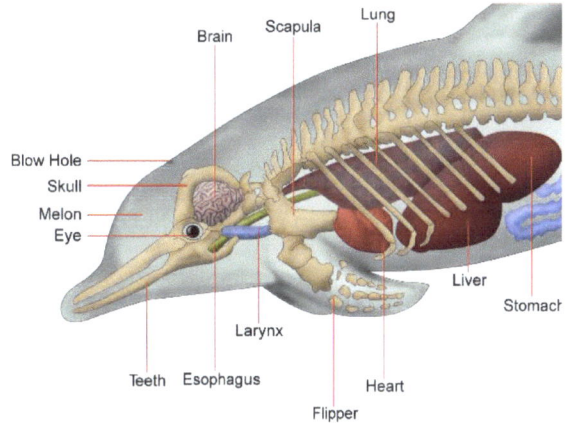

Fig. 7.2.5 The dolphin blowhole

Additionally, the Geologic Column (Chapter 2.2) is void of transitional forms. Most every species simply shows a variation of already extant forms or a species that is only aesthetically similar to an extant species. The fact of the matter is that no transitional form on the scale required by millions of years has ever been found.

Section 3
Dinosaur and Birds:

The textbook will say that dinosaurs are the ancestors to modern-day birds (Figure 7.3.1). The evidence for this is based entirely on small fragments of skin tissue.

Fig. 7.3.1 Dinosaur with feathers and without

The physical differences between dinosaurs and birds are far too numerous to list in this short work, but one major difference other than the absence of feathers is the bone structure.

Most birds are capable of flight because they have less density due to hollow bones. Terrestrial animals such as dinosaur and other lizards tend to have denser bones. As such a hollow bone structure can be a hindrance as it makes the animal much more fragile when it comes to fights or hunting for prey. Ground animals tend to have denser skeletons because that is a pattern that is much more opportunistic to their environment. Birds have hollow bones to aid in flight.

There are of course other differences between birds and reptiles (such as dinosaurs) but this will be sufficient for the present.

For many years scientists believed that dinosaurs and birds shared a common ancestor, but had little empirical evidence to work with until the proto-feather was "discovered". The proto-feather is a hypothetical tissue that is a transition between scales and feathers.

These so-called proto-feathers do not have many of the most important qualities that feathers require, such as branching. Besides, it offers no evolutionary advantage to only create part of a structure that does not do its job. In the evolutionary world view, if a trait does not increase the survivability and thus the ability to pass said genes on to the next generation, it will be diluted back into the gene pool. Therefore, there is no reason to suggest that the dinosaurs would have had feathers or proto-feathers.

Fig. 7.3.2 Birds found with dinosaurs

There are fully formed birds present in the fossil record in the same layer of the geologic column as dinosaurs. These include parrots, penguins, owls, sandpipers, albatross, flamingos, loons, ducks, cormorants, and avocets (Figure 7.3.2). These avians also appear just as they do today. According to the evolutionary world view, if two organisms exist in the same fossil layer, they lived in the same era. The unbiased scientist would conclude that these two organisms lived at about the same time.

Fig. 7.3.3 Skin of a plucked bird and a reptile

These birds not only represent a large variety but represent nine different orders of birds. All birds are classified under the class of aves. The order represents the next stage in the phylogenetic tree.

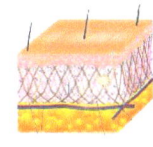

Fig. 7.3.4 Cross-section of human skin

As mentioned in Chapter 2.3, blood cells and other soft tissue have been found inside dinosaur bones. Recall also from Chapter 3.2 that skin imprints have been found as well. These skin imprints show clear indications of the type of skin these creatures had when alive. The evidence is clear and the skin looks reptilian and not avian (Figure 7.3.3).

If these are not feathers or even proto-feathers, the curious would want to know what they are. They may be protein fibers from the skin that were difficult for the microbes to break down (Figure 7.3.4). Microbes, like any other organism, can only digest certain materials. If a microbe came across a protein within the skin that it could not digest, the protein would remain just how it was until it was broken down due to natural decay or by incredibly slim odds, fossilized.

However, if dinosaurs are not the ancestors of birds, what do we know about these terrible lizards? More specifically, why were they so big?

Fig. 7.3.5 Reptiles and lizards

Many species of lizards never stop growing throughout their lives (Figure 7.3.5). A few other species of animals such as kangaroos (Figure 7.3.6) also share this trait. Fossils show very large animals and plants that lived in the past. What caused these plants and animals to grow so large is as of yet unknown.

Fig. 7.3.6 Kangaroo

For example, some species of dragonflies have been found that have a wingspan ranging from 65-70 cm. That is a little over two feet (Figure 7.3.7). Truly a very large creature that is scores bigger than anything that we see today.

Therefore, it does seem to logically follow that if the dragonfly can

Fig. 7.3.7 A comparison between the two varieties of dragonfly

grow to monumental proportions, why not lizards? If a simple dragonfly can grow to twice its size, how big can a lizard (which never stops growing) truly be?

Dinosaurs were just very large (possibly now extinct) species of lizards that grew exceptionally large. If so, dinosaurs are nothing more than big lizards (Figure 7.3.8).

Fig. 7.3.8 Large and Terrible Lizards

Chapter 7: Transitional Forms

Section 4
The History of Humans:

The textbook will also say that humans have a vast number of transitional forms (Figure 7.4.1). However, in reality, Darwinian scientists have falsified fossils to support their theory:

Fig. 7.4.1 A simplified depiction of supposed human ancestry.

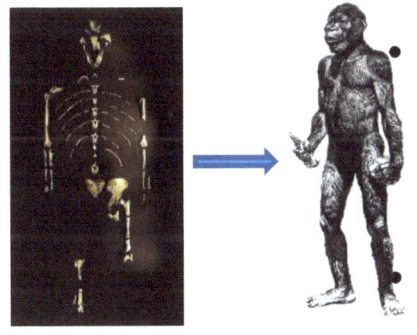
Fig. 7.4.2 Lucy has no hands or feet.

Lucy (*Australopithecus africanus*) (Figure 7.4.2) is depicted as having hands and feet in every display, but none were found.

Piltdown Man (*Eoanthropus*) was a forgery made from a human skull with the jaw of an orangutan grafted onto it.

Nebraska Man (*Hesperopithecus*) (Figure 7.4.3) and his wife were reconstructed from a single pig's tooth.

Fig. 7.4.3 Nebraska Man's Tooth

- Java Man (Pithecanthropus) was made from the skullcap of an ape, but a thigh bone was found 12 meters (40 feet) away. This was just assumed to be a part of the same creature as the skull cap.
- Orce Man (Figure 7.4.4) was a skull fragment from a four-month-old donkey.
- Peking Man (*Sinathropus*) was presented as an ape-man but is now known to be human.
- *Ramapithecus* was thought to be an ancestor of humans but is just an extinct species of orangutan.
- Neanderthal (Figure 7.4.5) was found to have died, according to Oxford, sometime around 1750. The textbook will say that this is an ancestor of modern man that lived several millennia in the past that started to stand erect. However, there is evidence to suggest that the reason it is bent over is a combination of disease (such as arthritis) and old age.

Fig. 7.4.4 Orce Man's skull cap

These are not just random examples but were actually posited by the scientific community until they were found to fakes. However, many evolutionists that still accept these fabricated fragments as fact.

There are other examples, but these serve to show how far evolutionists will go to support their hypothesis.

Fig. 7.4.5 Neanderthal skeleton

Many textbooks have stopped including the vast majority of these and other fabricated fossils, however, they will still be referenced to say that humans evolved from an ape-like ancestor without any proof or evidence.

There is also a discrepancy with how the skulls are reconstructed. Jack Cuozzo (Figure 7.4.6), an orthodontist, found that Neanderthal skulls were adjusted to support an evolutionary worldview. Cuozzo began examining skulls in the museums of Paris with a portable X-Ray. He found that there was a problematic issue with the jaw bone placement. If the jaw bone is attached further inward, it looks more human, but if the jawbone is attached further out, it looks more ape-like. What Cuozzo found was that the jawbones were often not attached to the skull. This means that the skulls in museums in Paris (and other displays around the world) were intentionally altered to support evolution by falsifying data.

Fig. 7.4.6 Jack Cuozzo

Throughout all recreations of supposed human ancestors, we also see a problem with artistic depictions. In most instances, if the artist wishes to depict an early transitional form, the artist will use dark brown skin.

While these examples are all falsified, some of the worst offenders were created by Ernst Haeckel in the late 1860s (Figure 7.4.7). Haekel was one of many scientists trying to prove Darwinian evolution. To do this, he decided to draw theoretical gestational changes in humans as compared to fish, salamanders, tortoises, chickens, hogs, calves, and rabbits. This was all done before reliable microscopes were available.

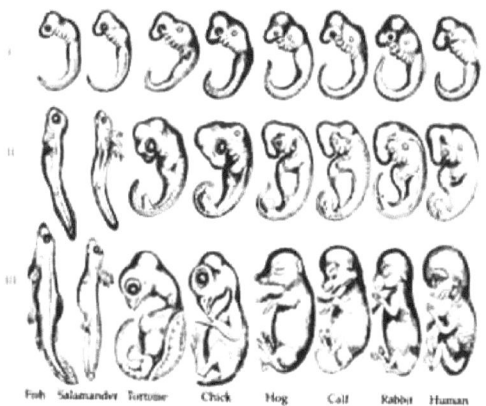

Fig. 7.4.7 Haekel's original drawings

His drawings indicated that the further down the evolutionary chain, the fewer physical similarities there were between the organisms in the womb or egg.

However, it was not long before it was found out that he faked the data (Figure 7.4.8). He was put on trial at his university. Even when approached with the frauds, he refused to admit his bias. He made the claim:

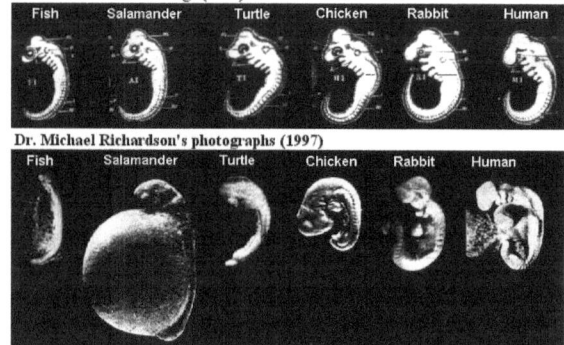

Figure 7.4.8 Photos of embryonic development

"A small percent of my embryonic drawings are forgeries; those namely, for which the observed material is so incomplete or insufficient as to fill in and reconstruct the missing links by hypothesis and comparative synthesis"

This is direct and intentional fraud. In any other field of science is someone published intentionally falsified data to support a conclusion, their ideas would never see the light of day.

Haekel justified his reasoning:

"I should feel utterly condemned...were it not that hundreds of the best observers, and biologists lie under the same charge."

This justification holds no merit and is factually untrue. The majority opinion of scientists does not make any hypothesis, theory, or law accurate or scientific. In the case of evolution, scientists who disagreed are silenced. Only those who agreed are allowed to have their opinions heard and published.

Even though Haekel was fired for his academic misconduct, his teachings and drawings are still used in textbooks today.

Section 5
Variation in Fossils

The textbook will say that we see organisms appear in the fossil record and that these show how they looked when they first moved across the Earth.

The only problem is that they look just as they do now.

For example, the octopus supposedly first evolved 250 million years ago (Figure 7.5.1).

However, fossils show little to no differences with extant (species that are still alive) varieties of octopus (Figure 7.5.2).

The same is true for virtually all species on Earth.

Thus there is no evidence found that indicates that animals have changed. Even if the two animals are different, this does not prove that one organism evolved from the other. The only thing that can be proven is that the two organisms are very similar, but have differences. It is entirely possible that the two organisms simply show to very distinct variations of the same species such as a Great Dane and a Chihuahua.

Fig. 7.5.1 An octopus fossil

Fig. 7.5.2 An extant octopus

Chapter 7: Transitional Forms

The textbook will outline how creatures have changed over the years by creating charts and graphics of very similar organisms together in order of increasing complexity (Figure 7.5.3).

This is based on the assumption that is central to evolution: Organisms slowly increase in complexity over the generations.

Even though there has never been any empirical indication of this, it will still be taught as fact. All that has ever been observed is a variation of already existing traits.

Therefore, no fossil evidence exists that shows transitional forms, just as there is no evidence of our hypothetical ancestors.

If the fossilized species is found to be different from the living species, the evolutionist will say that the fossil represents an earlier form of the creature. If the fossil is the same, the evolutionist will say that the creature remained static and this will also be evidence for millions of years.

Saying that there are transitional forms without any evidence is hardly scientific.

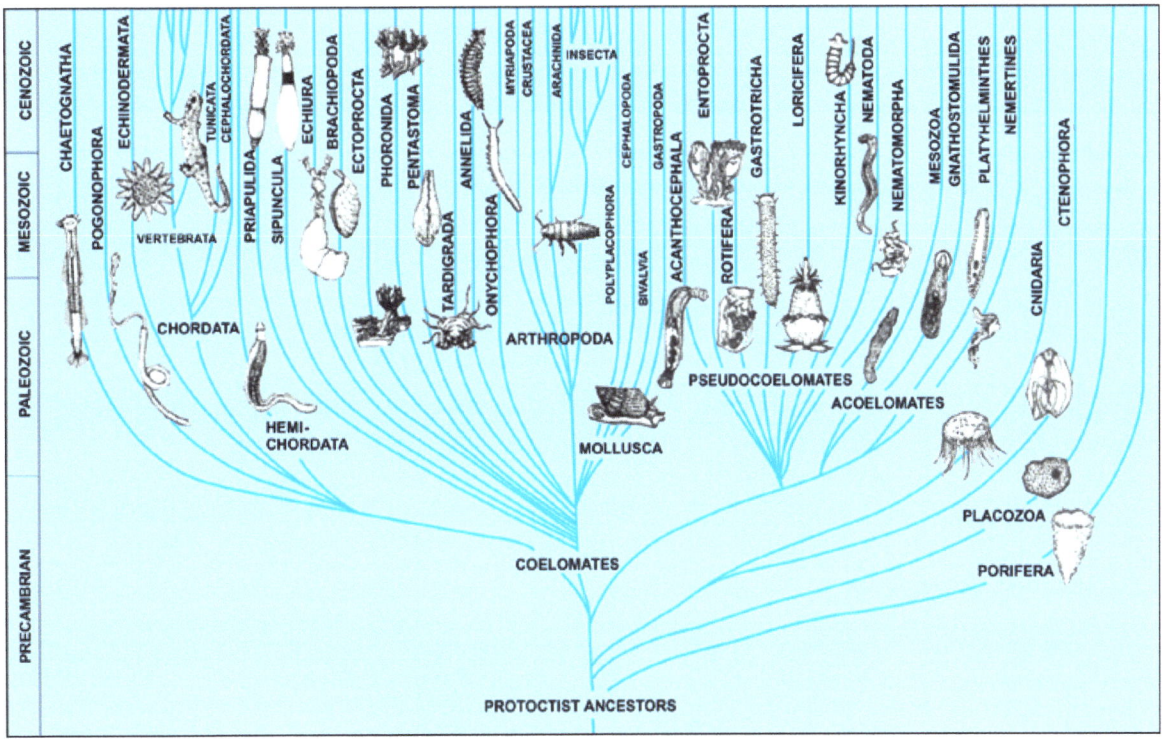

Fig. 7.5.3 A chart with hypothetical changes.

Section 6

Common Bones:

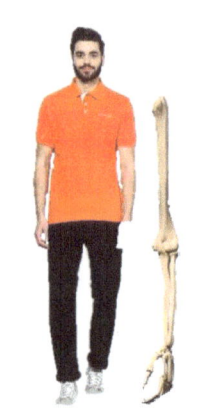

Fig. 7.6.1 Human's Arm

Fig. 7.6.3 Whale's Flipper

Fig. 7.6.2 Cat's Leg

Fig. 7.6.4 Bat's Wing

The human arm is made of several bones (Radius, Ulna, Humorous, Phalanges) (Figure 7.6.1). Very similar bones are also found in cats (Figure 7.5.2), whales (Figure 7.6.3), bats (Figure 7.6.4), and many other animals. There is no evidence that these are relatives on the tree of life, just that the bones are similar. Again, it bears repeating that the traits are not on analogous chromosomes (Chapter 6).

All this shows is that the bones have a very similar format and received the same name. This just proves that this is a very efficient bone pattern for a wide variety of functions.

This unique design can be used by humans for grasping, cats for walking, whales for swimming, and bats for flight. These many functions can all be accomplished by a very elegant system of bones.

Scientists gave the same name to these bones because of how similar they are to humans. These bones have very different functions but are similar in structure because it is a form that works.

Take for example a dresser, table, and television stand (Figure 7.5.5). Each one is made from the same basic component, wood. Despite their common element, they each perform different jobs, but this is not because of their common descent, but how they were put together.

Fig. 7.5.5 A dresser, table, and television stand.

Chapter 7: Transitional Forms

All three could hold a television set, but it would not be ideal to do your homework on a television stand. Nor would it be ideal to keep your clothes on a table.

If you walked into an old house and found a television sitting on a dresser rather than a television stand, you would not conclude that in the beginning televisions were set on dressers. You could however conclude that it was placed there because that is where the owner needed it. You would also not conclude that television stands and tables were developed later.

In the same way, just because these bones look similar does not prove that there is a common ancestor. This simply shows that many creatures have similar limbs because that is a system that works.

IN CONCLUSION:

The fabricated history of transitional forms is steeped in racial bias and quickly led to eugenics. (Section 1)

There is no empirical evidence that any aquatic mammal descended from a land-dwelling animal. (Section 2)

What we actually see are countless examples of fossils being falsified to support the evolutionary theory. Whether that be putting them together incorrectly, combining skeletons, or major guesswork. (Section 3)

The textbook will say that organisms have drastically changed over the years, but there is no data to support this. (Section 4)

Many other organisms across all kingdoms have very similar structures, but this offers no proof of a common ancestor. (Section 5)

There has never been a single transtional form on the scale needed for millions of years of evolutionary history found.

Questions for Further Discussion:

1. Why would scientists not discuss the undeniable racist history advocated by transitional forms? (Section 1)
2. Why would scientists say that the ancestors of whales and dolphins once walked on land without evidence? (Section 2)
3. If there is no evidence that dinosaurs ever had feathers, why would evolutionists posit that they did? (Section 3)
4. Why would scientists falsify skulls to make them fit with an evolutionary worldview? (Section 4)
5. Why would the textbook indicate an increase in physical complexity that is only based on guesswork? (Section 5)
6. If there is no evidence to support a position, what chaos would arise if researchers just made up facts to support their conclusions? (Section 5)
7. How are bones that perform vastly different functions evidence of a common ancestor? (Section 6)
8. Is there another hypothesis that will explain these phenomena?

Additional Reading

Secular Sources:

Britannica, The Editors of Encyclopaedia. "Peking Man." *Encyclopædia Britannica*, Encyclopædia Britannica, Inc., 24 Mar. 2009, www.britannica.com/topic/Peking-man.

Britannica, The Editors of Encyclopaedia. "Sivapithecus." *Encyclopædia Britannica*, Encyclopædia Britannica, Inc., 6 May 2014, www.britannica.com/topic/Sivapithecus.

Gould, Stephen Jay. *Ontogeny and Phylogeny*. Belknap Press of Harvard University Press, 2003.

J.G.M. Thewissen, E.M. Williams, L.J. Roe, and S.T. Hussain, Skeletons of terrestrial cetaceans and the relationship of whales to artiodactyls, *Nature* 413:277–281, 20 Sept. 2001.

"Orce Man." *Xu Bing: Square Word Calligraphy Classroom - The Miriam and Ira D. Wallach Art Gallery*, Columbia University. Libraries. Digital Program Division, www.columbia.edu/itc/anthropology/v1007/castro/tsld006.htm.

P.D. Gingerich, N.A. Wells, D.E. Russell, and S.M.I. Shah, *Science* **220**(4595):403–6, 22 April 1983; P.D. Gingerich, *Journal of Geological Education*. **31**:140–144, 1983.

"Neanderthals." *Evolution of Modern Humans: Early Modern Human Culture*, Palomar College, www2.palomar.edu/anthro/homo2/mod_homo_2.htm.

Pakicetus ... eight years on. Illustration: Carl Buell http://www.neoucom.edu/Depts/Anat/Pakicetid.html

"Pakicetus Spp." *Rodhocetus Spp. | College of Osteopathic Medicine | NYIT*, New York Institute of Technology, www.nyit.edu/medicine/pakicetus_spp/.

"Piltdown Man." *When Did Dinosaurs Live? - Natural History Museum*, Natural History Museum, www.nhm.ac.uk/our-science/departments-and-staff/library-and-archives/collections/piltdown-man.html.

Reproductive Isolation, Berkeley University, evolution.berkeley.edu/evolibrary/article/0_0_0/lines_03.

Reynolds, Eileen. "Ota Benga, Captive: The Man the Bronx Zoo Kept in a Cage." *NYU*, New York University, 7 Aug. 2015, www.nyu.edu/about/news-publications/news/2015/august/pamela-newkirk-on-ota-benga-at-the-bronx-zoo.html.

Richardson, M. K., et al. "There Is No Highly Conserved Embryonic Stage in the Vertebrates: Implications for Current Theories of Evolution and Development." *Anatomy and Embryology*, vol. 196, no. 2, 1997, pp. 91–106., doi:10.1007/s004290050082.

Wilford, John Noble. "ABORIGINE EVOLUTION ASSUMES A MAJOR ROLE." *The New York Times*, The New York Times, 24 July 1984, www.nytimes.com/1984/07/24/science/aborigine-evolution-assumes-a-major-role.html.

Additional Reading

Religious Sources:

Batten, Dan. "Modern Birds Found With Dinosaurs." *Creation.com | Creation Ministries International*, Creation Ministries International, creation.com/modern-birds-with-dinosaurs.

Cuozzo, Jack. *Buried Alive: the Startling Truth about Neanderthal Man*. Master Books, 1998.

Ham, Ken. "Are Dinosaurs Related to Birds?" *Answers in Genesis*, Answers in Genesis, 27 Jan. 2011, answersingenesis.org/kids/dinosaurs/are-dinosaurs-related-to-birds/.

Henderson, Doug. "Bringing Lucy to Life." *Answers in Genesis*, 1 Jan. 2013, answersingenesis.org/human-evolution/lucy/bringing-lucy-to-life/.

Lawwell, Stephen. "Echoes of Eden." *Evolution and Abortion*, Echoes of Eden, 21 July 2014, www.echoesofeden.org/articles/abortion/evolution-abortion.

Menton, David. "Did Dinosaurs Evolve into Birds?" Answers in Genesis, Answers in Genesis, 7 Sept. 2018, answersingenesis.org/dinosaurs/feathers/did-dinosaurs-evolve-into-birds/.

Mortenson, Terry. "National Geographic Is Wrong and so Was Darwin." *Answers in Genesis*, 6 Nov. 2004, answersingenesis.org/charles-darwin/national-geographic-is-wrong-and-so-was-darwin/.

Taylor, Ian T. "'Nebraska Man' Revisited." *Answers in Genesis*, 1 Sept. 1991, answersingenesis.org/human-evolution/piltdown-man/nebraska-man-revisited/.

Who Was 'Java Man'?" *Creation*, vol. 13, no. 3, June 1991, pp. 22–23.

Chapter 9: Starlight

NOTES:

CHAPTER 8
Vestigial Organs

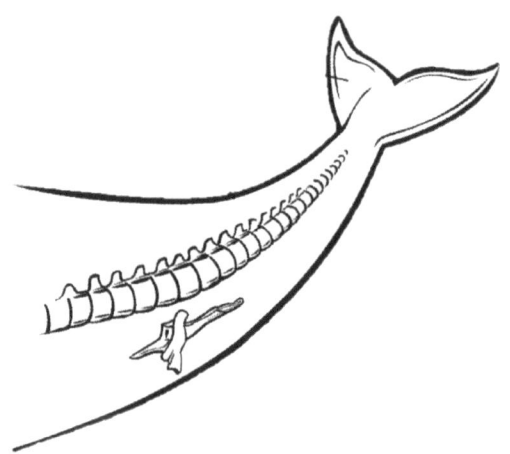

Chapter 8: Vestigial Organs

What Does the Textbook Teach?

The textbook is going to teach that humans and other organisms have structures that were useful in the past. They remain, but no longer serve a function. They posit that this is evidence of millions of years of evolution. (Section 1 & Section 2)

It will say that there are portions of the human genome that are leftovers from our ancestors down the evolutionary chain (Section 3).

Finally, the textbook will say that there are organs that evolve away in the womb. (Section 4)

Section 1
Vestigial Organs in Animals:

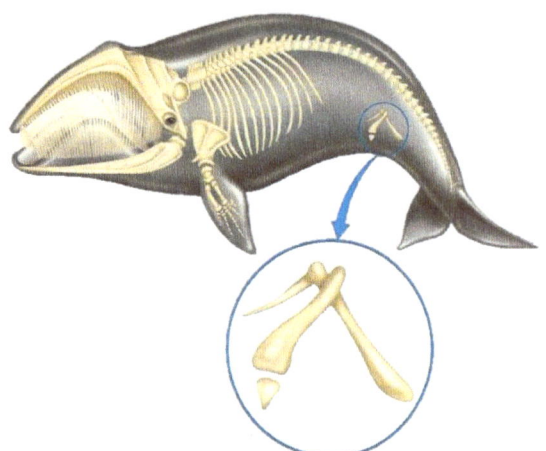

Fig. 8.1.1 The whale's bones

Fig. 8.1.2 The snake's bones

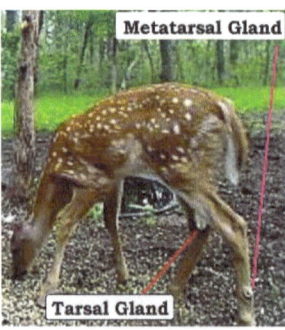

Fig. 8.1.4 Deer scent glands

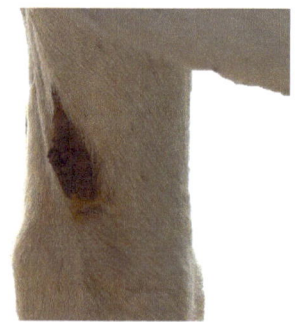

Fig. 8.1.3 A horse chestnut.

A vestigial organ is a structure that no longer serves a purpose. The textbook will say that it might have served a purpose in one of the ancestors of the organism countless years ago. Since evolution eliminates only that which is harmful or adds that which is helpful if a structure is neither harmful nor helpful, it will remain unchanged. This becomes a vestigial structure.

According to the textbook, it will serve no function because it is no longer vital.

The textbook will say that one of the most important vestigial organs are the remnants of legs in whales and snakes.

If snakes and whales once walked around, they must have had legs. If they once had legs and no longer do, they must have lost the legs.

The skeletons of both creatures (Figures 8.1.1 and 8.1.2) have shown that they have both a pelvis and very small "leg" bones. The only problem is that these structures were never used to walk on land.

These cannot be vestigial organs because they still serve an essential function in reproduction. There is no evidence to suggest that these "leg" bones could have ever served any other purpose.

Even if these were legs, this is evidence of losing information, not gaining information.

Another supposed vestigial organ is the horse chestnut (Figure 8.1.3). These are small structures on the inside of the front legs of horses.

The textbook will say that they are vestiges of toes. If this is true, this is a reshuffling of already present genetic information.

Furthermore, these may be scent glands such as those found in deer (Figure 8.1.4). They can also be used to identify and distinguish between horses in the same way that fingerprints can identify people.

Section 2
Vestigial Organs in Humans:

The textbook will list many examples of supposed vestigial structures and say that since these do not have a function, humanity is evolving them away. Some of these structures are not vital to life. A person could live without them. In the same way, a person could live without eyes, but that does not mean that they do not serve a unique and important purpose.

As scientists learn more about anatomy, they learn that structures once believed to be useless do serve a purpose:

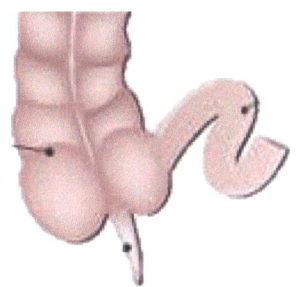

Fig. 8.2.1 The appendix

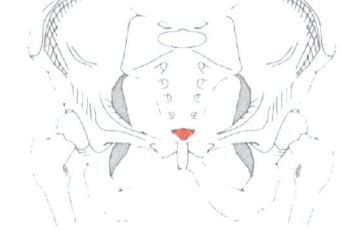

Fig. 8.2.2 The coccyx

- The appendix (Figure 8.2.1) harbors good bacteria and helps to fight dangerous bacteria. Supposedly our ancestors had a different diet. The appendix was supposedly an extension of the intestine from when our ancestors were solely herbivorous. Animals that subsist entirely on plants have a much longer intestinal tract to break down plant matter.
- The coccyx (Figure 8.2.2) helps maintain balance in a sitting position. It also serves a vital role in keeping the internal organs within the body. This structure supposedly connected our tails to the rest of the body. Since humans do not have tails, there is theoretically no need for this bone, but it still serves a vital function.
- Wisdom teeth (Figure 8.2.3) have a purpose in chewing and very few even need to be removed because most will not cause medical problems. In theory, it remains because our diet has changed over the years. A wisdom tooth can be removed because the tissue around it got infected, but this does not mean that it is vestigial. Rather, it means that the structure was in the wrong place at the wrong time.

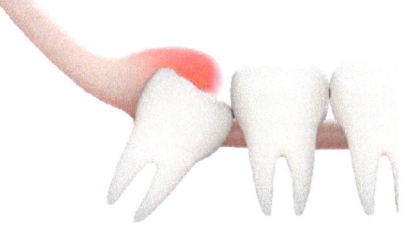

Fig. 8.2.3 The wisdom teeth

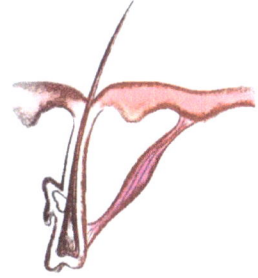

Fig. 8.2.4 The erector pili

- Erector Pili (Figure 8.2.4) and body hair serve as insulation from the cold and serve a sensory function. Goosebumps occur when the body is subjected to extreme cold or fear. This causes the hair to create pockets of air to help warm the body.

These structures and organs have been posited by scientists to be vestigial remains of our ancestors and thus have no purpose. There are many structures across the body and scientists do not fully understand all of them. There are many more organs that have a purpose but are considered vestigial. These overlooked organs are believed by scientists to be vestigial remains of our evolutionary past and thus are irrelevant.

Imagine ripping out the pages (Figure 8.2.5) to an encyclopedia that you believed were irrelevant to your studies. At some point, you may need the pages that you have torn out. In the same way, the countless organs that are considered vestigial, do have a purpose, even if scientists do not fully understand it.

In fact, in the year 1959 there were about 1.4 million tonsillectomies, but only 500,000 in 1979 and then about 250,000 in the 2010 decade. Why was there such a drop? It might have to do with the fact that scientists began finding a use for tonsils and that they were not just useless vestiges.

Fig. 8.2.5 Pages being torn out.

Chapter 8: Vestigial Organs

Section 3
Junk DNA:

The genetic code that makes up any organism is incredibly intricate.

The textbook will say that there are large sections of DNA that serve no purpose. This "functionless" DNA is present because of how far evolved we are.

Simply put, the textbook will say that this genetic material is nothing more than remnants of our ancestors that no longer serves a purpose.

This is false. Firstly, just because we do not understand the purpose of a structure does not mean that it is useless. Secondly, many pieces of so-called junk DNA do serve a purpose.

Telomeres (Figure 8.3.1) are a fine example. These are long structures that exist on the ends of the DNA strand. Each time the DNA is copied, a small part at the end is lost. The telomeres are very long and can last a lifetime. If the telomeres were not present, each time a cell divided, it would start cutting into the coding material of the DNA. This would very soon if not immediately be fatal.

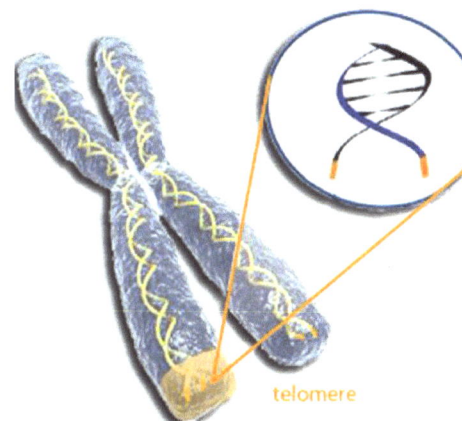

Fig. 8.3.1 A telomere

Another structure is non-coding portions of DNA that exist between coding portions. These sections of DNA do not code for any enzyme or protein and the cell cannot do a single thing with them. This does not mean that they do not have a purpose. Many of these sequences act as punctuation between coding messages. They are the molecular equivalent of periods and spaces so that the cell can keep the messages separate. Imagine reading a paragraph with no spaces or punctuation (Figure 8.3.2). It is possible, but it is much more difficult.

There are other portions of the DNA that control how the DNA is copied and coded.

Some genes are simply repeats of genes that are present in other places on the genome. Supposedly

HUMPTYDUMPTYSATONAWALLHUMPTYDUMPTYHADAGREATFALLALLTHEKINGSHORSESANDALLTHEKINGSMENCOULDNTPUTHUMPTYTOGETHERAGAIN

Fig. 8.3.2 A paragraph without punctuation.

these are there because the same trait evolved twice. However, these serve as backup genes should the others be deleted by a copying error or by environmental factors such as UV light.

The genetic code is phenomenally complex and scientists have yet to fully understand everything present within, but just because we do not understand it, does not mean that it is useless.

Section 4
Non-Existent Organs:

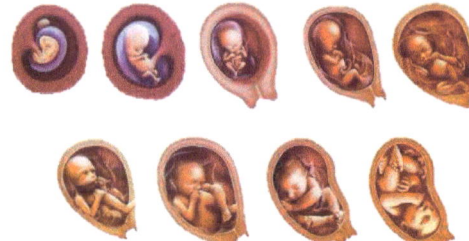

Fig. 8.4.1 The development of a human.

The textbook will say that humans develop many organs in the womb (Figure 8.4.1) that are not present after birth.

These are remnants of our evolutionary tree such as gill slits, webbed digits (fingers and toes), and tails.

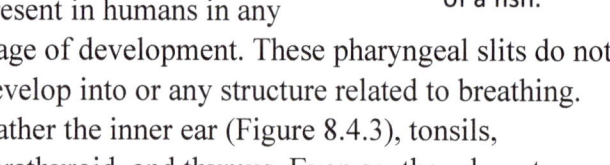

Fig. 8.4.2 The gills of a fish.

Fish use gills to breathe (Figure 8.4.2). The textbook will say that humans have pharyngeal slits like those of fish while in the womb that are leftover from when our ancestors were fish.

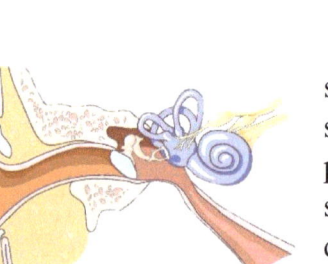

Fig. 8.4.3 Internal ear structure

There is no scientific evidence to even suggest that gills are present in humans in any stage of development. These pharyngeal slits do not develop into or any structure related to breathing. Rather the inner ear (Figure 8.4.3), tonsils, parathyroid, and thymus. Even so, they do not even appear on analogous chromosomes (see Chapter 6).

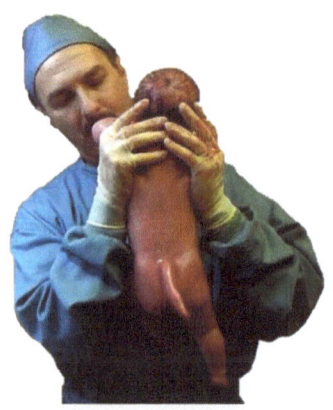

Fig. 8.4.4 A human born with a "tail".

The tail (Figure 8.4.4) in question is unconnected to the spine. Without this pivotal connection, it cannot function as a tail. This is more than likely, just a growth of tissue and growth of an organ is more common than expected. Some people are born with three kidneys and even with horn-like tissue (Figure 8.4.5).

Fig. 8.4.5 A horn growing on a woman

The webbed digits (Figure 8.4.6) come when the fingers are first being formed. Rather than actually growing out of the hand, the fingers are "sculpted" during embryonic development. Webbed digits merely indicates that the job was unfinished before the child was born.

Chapter 8: Vestigial Organs

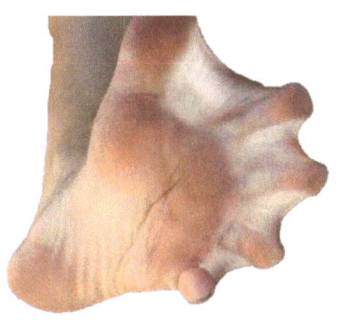

Fig. 8.4.6 Webbed feet

The textbook will say that these structures are remnants of our evolutionary past. However, a more logical explanation is that they are all simply premature components of humans that have no connection to our supposed ancestors.

IN CONCLUSION:

Scientists have found a vital purpose for virtually every organ in all animals studied. Each organ has a function and a purpose, even if we do not know what it is. (Section 1)

The same is true for humans. Researchers once considered many structures vestigial, however, they are now found to be important. (Section 2)

There is no evidence of junk DNA. Just because we do not know the purpose of each component does not make it useless. As we learn more about the most complicated molecule, we see a purpose that is unrelated to evolution. (Section 3)

Humans have many organs that are similar to other animals, but this does not show ancestry. Contrary to what many scientists will say, tails and webbed feet have no connection to past stages of evolution. (Section 4)

Questions for Further Discussion:

1. Since the vestigial bones in whales and snakes are used for reproduction, why would scientists say that they are remnants of legs? (Section 1)
2. Why would scientists prematurely say that a structure has no function before doing thorough research? (Section 1 & Section 2)
3. What might happen if a surgeon deemed a useful organ useless or vestigial? (Section 2)
4. What would happen, long term, to research if we treated something of importance as just tissue? (Section 2)
5. Give examples of things that were thought to have had no function in the past (vestigial), but now are known to have a function (essential)? (Section 2)
6. Why would scientists falsely claim there is junk DNA? (Section 3)
7. Why would an organ be called vestigial when humans do not even have the organ in question? (Section 4)
8. Is there another hypothesis that will explain these phenomena?

Chapter 8: Vestigial Organs

Additional Reading

Secular Sources:

Care, AllKids Urgent. "The History of Tonsil Removal." *AllKids Urgent Care*, AllKids Urgent Care, 19 Feb. 2015, mysickkid.com/tonsils-gilbert-pediatric-urgent-care/.

Hadden, Will A., III, D.V.M., *Horseman's Veterinary Encyclopedia*, The Lyons Press, Guilford, Connecticut, 2005, page 169.

Lu, Weisi, et al. *Telomeres - Structure, Function, and Regulation*. National Center for Biology Information, 21 Sept. 2012, www.ncbi.nlm.nih.gov/pmc/articles/PMC4051234/.

Martin, Loren G. "What Is the Function of the Human Appendix? Did It Once Have a Purpose That Has since Been Lost?" *Scientific American*, www.scientificamerican.com/article/what-is-the-function-of-the-human-appendix-did-it-once-have-a-purpose-that-has-since-been-lost/.

Thompson, Helen. "Whale Reproduction: It's All in the Hips." *USC News*, University of Southern California, 26 July 2017, news.usc.edu/68144/whale-reproduction-its-all-in-the-hips/.

Wellness, Berkeley. "Coccyx (Tailbone) Pain: Causes, How to Treat." *@Berkeleywellness*, Berkeley University, www.berkeleywellness.com/self-care/preventive-care/article/pain-coccyx.

Religious Sources:

Mitchell, Tommy, and Elizabeth Mitchell. "Something Fishy About Gill Slits!" *Answers in Genesis*, 14 Mar. 2007, answersingenesis.org/evidence-against-evolution/something-fishy-about-gill-slits/.

Wolfrom, Glen, Horse Chestnuts, *Creation Matters* **3**(4):5, July–August 1998.

NOTES:

Chapter 9: Starlight

NOTES:

CHAPTER 9
Starlight

Chapter 9: Starlight

What Does the Textbook Teach?

      The textbook will say that the stars are unbelievably far away. (Section 1 & Section 2)

    It will also say that new stars are always forming and we see the remains of dead stars. (Section 3 & Section 4)

Section 1
Measuring Stars:

Before delving into this, it is important to learn why scientists care that the astral bodies are millions of light years away. If the stars are millions of years old, the universe therefore must be millions of years old as well. Therefore, the Earth and life may be millions of years old also.

Surveyors and scientists sometimes have a very difficult time concluding where objects are because they cannot directly measure them.

However, because the laws of mathematics are set in stone, if you know enough about angles and distance, it is simple to calculate exactly how far away an object is by using a theodolite and parallax trigonometry (Figure 9.1.1).

Fig. 9.1.1 A theodolite used for surveying.

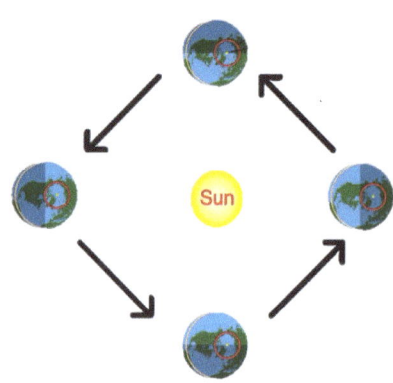

Fig. 9.1.2 Earth's orbit around the sun.

Scientists measure the angle of a star in relation to the Earth on a certain day. Then they take the same measurement again six months later when the Earth is on the other side of its orbit (Figure 9.1.2).

Proxima Centauri is thought to be one of the closest stars to Earth (Figure 9.1.3). Using this trigonometric method, the textbook will say that this star is about 4.22 light-years away. That is about 24,807,800,000,000 miles or 24.8 trillion miles.

Fig. 9.1.3 Proxima Centauri

The distance between Earth at measurement one and measurement two is about 185,911,614 miles or 185.9 million miles.

If we were to put this as an isosceles triangle, this would put the two sides at 24.8 trillion miles and the base of the triangle at 185.9 million miles (Figure 9.1.4).

Now that we have that information, we can make a relative comparison. If two surveyors were one inch apart, they would both be looking at an object a little over 25 miles away. That is the distance of 813 Olympic swimming pools or 445 football fields away.

This distance creates a huge margin for error. It is unimaginable that anyone could get even a slightly accurate number with such skewed information.

Even if the light was not bent to make the star appear elsewhere, the star still has to be seen through our very thick atmosphere.

Fig. 9.1.4 The isosceles triangle (not to scale)

Chapter 9: Starlight

Even if there was an accurate image of the star through the atmosphere, we are on a moving planet. The Earth is rotating on its axis at about 40,070 Km/hr (24,898 MPH). The Earth is also revolving around the sun at 107,000 Km/hr (67,000 mph). The solar system is even moving within the Milky Way at 720,000 Km/Hr (448,000 mph). Then the entire galaxy is moving at 112Km/Sec (70 Mi/Sec). Each of the other spatial bodies are also moving at similar speeds. This is the equivalent of getting on a tilt-a-whirl at the fair and trying to see how close someone on the other side of the park is to the front gate.

The space mission Gaia mapped 1 million stars just in the Milky Way Galaxy and there may be anywhere from 100-200 billion galaxies in the universe. This means that there should be at least 100 quadrillion (100,000,000,000,000,000) stars in the universe. With so many stars, we should be able to see stars in every stage of development, but we do not. All that has been seen is a spot of light that gets brighter.

However, trigonometry is not practical with distances greater than 100 light years. In order to measure these distances outside of trigonometry, astronomers must employ even more complicated methods.

Scientists use the color of the star and how bright it is to calculate the temperature and how far away it is, even though there is no way to confirm the calculations with empirical data.

Therefore, it is unquestionable that due to the large margin for error, these stellar measurements should not be taken at face value.

Section 2
Locating Stars:

Fig. 9.2.1 Albert Einstein

Since we now know that it is near impossible to tell how far away a star is, we need to look at how we know where it is.

Albert Einstein (Figure 9.2.1) was famous for many of his discoveries in physics, but one of his more spectacular theories is that light bends in response to gravity. His theory has been proven correct by a truly elegant experiment.

During a solar eclipse (Figure 9.2.2), scientists took as many pictures of the eclipse as possible. They noticed that starlight immediately next to the sun appeared to move between the pictures (Figure 9.2.3).

Fig. 9.2.2 An eclipse of the sun.

This showed that the light that was reaching Earth from other stars was bending in response to our Sun's gravity.

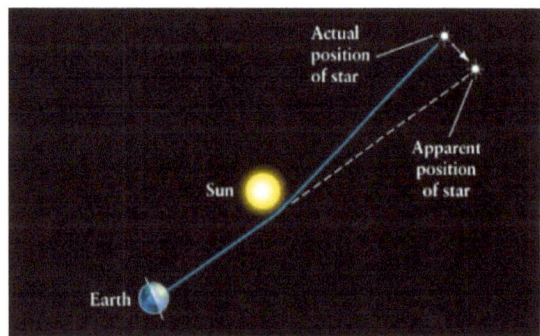

Fig. 9.2.3 Stars move

Fig. 9.2.4 Bodies moving away.

As a result, it is more difficult to know exactly where a star is as countless bodies could be in the way to artificially distort its location. Even evolutionists realize that beyond about 100 light years, the distance is too vast and the angle too small to get any practical data.

Further complicating these calculations is the redshift. The redshift is a property that is expected from the Big Bang hypothesis. The textbook will say that bodies move away from the initial point of expansion at the center of the universe (Figure 9.2.4).

We only see objects that are moving away from our frame of reference. This creates an effect known as a redshift. Therefore, it is believed the universe is expanding in relation to redshift.

Fig. 9.2.5 Galaxy NGC 4319 & Quasar Markarian 205

However, measurements and calculated expansion rates from redshift are purely hypothetical, when practical (experimental) science is applied, the story is much different.

Galaxy NGC 4319 and the Quasar Markarian 205 (Figure 9.2.5) are at least according to redshift hypothesis, billions of years apart. However, they are connected by a luminous bridge that should not exist.

If redshift theory calculations were true, we would not see this bridge.

Earlier, we mentioned light-years in regards to distance. A light-year is the distance that light can travel in one year. However, scientists have been able to alter the speed of light in a laboratory.

Physicists were able to control conditions to slow a beam of light down to 38 MPH, a speed slow enough to see with the naked eye (Figure 9.2.6).

Scientists have also been able to create circumstances in which light moves 300 times faster than the hypothetical speed of light.

These were all done in a lab, but is there any evidence that the speed of light changes in nature?

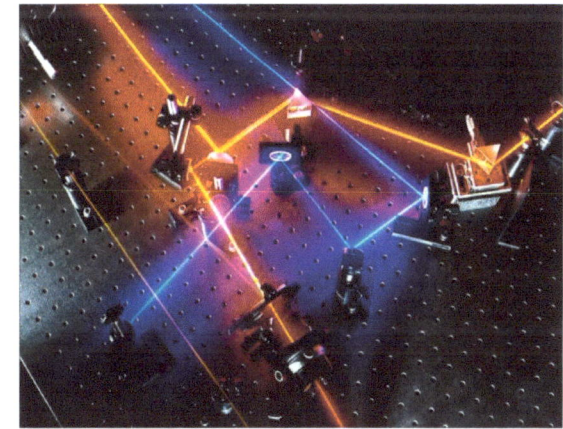

Fig. 9.2.6 Laser bouncing off mirrors.

The universe is hypothetically calculated to be 13.8 billion years old. However, some scientists believe that the universe may be as much as 7 trillion miles across. This means that even if the universe expanded from a singularity (see Section 5.2), this means that the speed of light would have to have been faster in the past to account for this distance. This means that the accelerated motion would have had to take place with all of the mass of the early universe which came from an infinitely dense sphere. This is impossible even if gravity did not slow the process down substantially, because this world view claims that the speed of light is a standard metric.

In short, the speed of light is not constant. Therefore, the techniques used to learn where the star is, measure the distance to the star, the unit used to measure the distance, and even the hypothetical movement of the star are all defective. As a result, how can any stellar measurement be considered scientific? If it is not scientific, how can they used as an accurate gauge for the age of the universe?

Section 3
Birth of Stars:

Recall from Chapter 5.1 that stars are supposedly formed from gasses that are condensing over long periods of time.

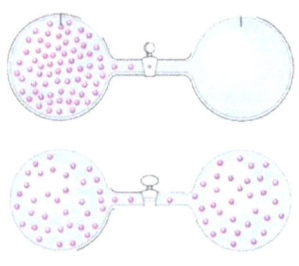

Fig. 9.3.1 Gas expands in a vacuum.

How the stars first formed has been an ongoing mystery.

The textbook will say that the stars are formed from condensing matter. How this reaction occurs in the vacuum of space, when super-hot gasses spread out and expand by their very nature is never explained (Figure 9.3.1).

Recall that the red shift from the previous section says that all spatial bodies are moving away from the initial point of expansion. Despite this, the stars are still forming while everything moves from the central point.

To support this theory of star formation, scientists expect to find so-called "Stellar Nurseries". One such nursery is (supposedly) found in the Elephant's Trunk Nebula (Figure 9.3.2).

However, there is no proof that the nebula is forging the stars. Rather, they may simply be coming into view. Thus, the scientists observing this data simply assume that the stars are forming.

Nebulas are large clouds of dust and gas. Thus if the dust clears, we would be able to see the stars coming into view, just like sunlight that streams through the clouds on Earth.

In short, there is little data to support that new stars are forming today.

Either way, if the universe is billions of years old, we should see countless stars in every stage of development in every direction.

Recall from the previous section, that we have a difficult enough time locating where adult stars are, let alone where the stars are when they are forming. Recall that one of the problems with locating the stars is that the light is disrupted, yet a nebula would very drastically distort the light. Still, we cannot even be sure that the young star is located in the nebula at all.

Fig. 9.3.2 The Elephant Trunk Nebula

Section 4
Death of Stars:

A violently exploding star is known as a supernova. When this occurs, it creates a Supernova Remnant (SNR) (Figure 9.4.1) such as one found in the Crab Nebula (Figure 9.4.2). This reaction was visible from Earth in the year 1054 AD.

Fig. 9.4.1 A SNR

The SNR should reach 300 light-years after 120,000 years. Thus, if the millions of years theory is correct, we should expect to see countless SNRs today.

Supernovas have three stages. If the universe were billions of years old, scientists would expect to see about two in the first stage, 2,260 in the second stage, and 5,000 in the third stage.

Fig. 9.4.2 The Crab Nebula

In actuality, what we see is five in the first stage, 200 in the second stage, and none in the third stage. This is not what we would expect from a universe that is billions of years old.

To put the same numbers another way, there is a supernovae explosion in the Milky Way Galaxy every 25 years on average, but we do not see enough for millions of years of evolution. There are only enough supernovae for a few thousand years.

These stellar remains imply a very different age of the universe.

Section 5
Forming Elements:

Once more, it bears repeating that there is a problem with stellar formation. The stars are needed to form the elements, but the elements are also needed to form the stars (see Chapter 5).

The more complex elements are brought about by smaller elements fusing at amazing rates.

This does not touch on how the first hydrogen atom was formed...

However, the same stars were made by these complex elements (Figure 9.5.1).

Fig. 9.5.1 The chicken and the egg.

Chapter 9: Starlight

IN CONCLUSION:

In actuality, what we see is that the techniques used to measure the distance to stars is severely flawed. (Section 1 & Section 2)

There is no evidence to suggest that stars formed slowly. Furthermore, if they did form slowly, we should see many more dead stars. (Section 3 & Section 4)

Stars are needed to form elements and the elements are needed to form the stars. (Section 5)

Since the methods used for determining the distance to stars is unscientific, they should not be used to determine the age of the universe or to posit on the age of the Earth.

Questions for Further Discussion:

1. Is it possible that the stars are not as far away as we think? (Section 1 & Section 2)
2. Why is redshift still taught if the physical observations do not support the hypothetical calculations? (Section 2)
3. Why is a lightyear considered a standard measurement? (Section 2)
4. Why do we not see many more stars in every stage of development? (Section 3)
5. Why are there so few dead stars if the universe is billions of years old? (Section 4)
6. Which came first, the stars or the elements? (Section 5)
7. Is there another hypothesis that will explain these phenomena?

Additional Reading

Secular Sources:

Baraniuk, Chris. "Earth - It Took Centuries, but We Now Know the Size of the Universe." *BBC*, BBC, 13 June 2016, www.bbc.com/earth/story/20160610-it-took-centuries-but-we-now-know-the-size-of-the-universe.

Cromie, William J. "Physicists Slow Speed of Light." *Harvard Gazette*, Harvard Gazette, 23 Feb. 2018, news.harvard.edu/gazette/story/1999/02/physicists-slow-speed-of-light/.

Kelley, Peter. "New Discovery Proxima b Is in Host Star's Habitable Zone - but Could It Really Be Habitable?" *Office of Minority Affairs Diversity*, University of Washington, 29 Aug. 2016, www.washington.edu/news/2016/08/29/new-discovery-proxima-b-is-in-host-stars-habitable-zone-but-could-it-really-be-habitable/.

"Illuminating Relativity: Experimenting with the Stars." *Astrology: Is It Scientific?*, Berkeley University, undsci.berkeley.edu/article/natural_experiments.

"Measuring Distances to Stars." *Methods of Observational Astronomy*, University of California, earthguide.ucsd.edu/virtualmuseum/ita/06_3.shtml.

Redd, Nola Taylor. "How Big Is the Universe?" *Space.com*, Space, 7 June 2017, www.space.com/24073-how-big-is-the-universe.html.

"Stellar Nursery." Edited by NASA Content Administrator, *NASA*, NASA, 23 Mar. 2015, www.nasa.gov/multimedia/imagegallery/image_feature_643.html.

Than, Ker. "Scientists Mess with the Speed of Light." *LiveScience*, 19 Aug. 2005, www.livescience.com/396-scientists-mess-speed-light.html.

Whitehouse, David. "SCI/TECH | Beam Smashes Light Barrier." *BBC News*, BBC, 19 July 2000, news.bbc.co.uk/2/hi/science/nature/841690.stm.

Additional Reading

Religious Sources:

Sarfati, Johnathan. "Exploding Stars Point to a Young Universe." *Exploding Stars Point to a Young Universe*, vol. 19, no. 3, June 1997, pp. 46–48.

Snelling, Andrew. "Galaxy-Quasar 'Connection' Defies Explanation." *Journal of Creation*, vol. 11, no. 3, Dec. 1997, pp. 254–255.

NOTES:

Chapter 9: Starlight

NOTES:

Epilogue

This work has discussed many of the supposed pieces of evidence for evolution and millions of years. Every piece of evidence presented has been given fair treatment and discussed factually.

As you continue to expand your mind, exploring both school and the real world, know that you will never meet a single person that does not have some bias.

These biases cloud our judgment and lead us to false beliefs that we would normally reject. Therefore, when examining any subject matter it is vital to first examine your own biases and how they cloud your judgment.

In Chapter 1, we discussed the scientific method of how to turn a hypothesis into a theory or law. We have tested the hypothesis of evolution throughout this work, but it has yet to be proven true. Therefore, it remains just that: A hypothesis.

Hence, evolution is neither a theory nor a law after each chapter, you were asked if there was another explanation for the phenomena described. In short, there is no other "scientific" explanation.

However, teaching something that is factually untrue, just because there is no other "scientific" explanation, is immoral.

Sherlock Holmes, the master of deductive reasoning, once said:

"When you have eliminated the impossible, whatever remains, however improbable, must be the truth."

Let us for a brief moment examine all of the things that must be true for millions of years of evolution to be true:

- The tectonic plates need to show evidence of large-scale changes over a long time...
- The geologic column has to accurate...
- Fossils cannot extend through multiple rock layers...
- Fossils had to form slowly...
- There must be no evidence that dinosaurs ever lived with man...
- Cells must be able to form from purely natural processes...
- Radioactive dating needs to be reliable...
- The universe needs to have formed by expansion...
- All life must have a common ancestor...
- Genes and entire chromosomes need to appear from nothing...
- All organisms must have transitional forms...
- Organisms must show traces of their evolutionary past in the womb or egg...
- All organisms must have vestigial structures...
- Junk DNA must be present in the genome...
- Humans must have organs that disappear in the womb from our ancestors...
- The distance to stars must be able to be calculated...
- The redshift must be a standard metric for the movement of stellar bodies...
- Stars must be able to be formed by purely natural processes...

- We should see many more dead stars…

Since none of these are supported by empirical and scientific data, evolution MUST remain a hypothesis.

Perhaps we should take a page from the book of Mr. Holmes.

I love science and I love to learn. Putting false explanations in a book just because another hypothesis is inconvenient or unpopular is almost criminal.

A final question to discuss:
Why would any alternative view, notion, hypothesis, or thought be completely ignored and rejected by the scientific and academic communities?

Glossary

Amino Acid-a molecule that is the main component of proteins

Analogous-things which are associated with each other because of how similar they are in form or function

Anthropology-the branch of science that deals with people groups and culture

Antibiotic Resistance-the occurrence of a microbe gaining immunity to a drug

Animal-a multi-cellular organism with a membrane-bound nucleus

Appendix-a small organ at the end of the small intestine that helps regulate bacteria

Astronomy-the branch of science that deals with stars and other celestial bodies

Atom-the smallest component of an element

Ave-the class that represents all birds

Big Bang Hypothesis-a hypothesis that posits the universe began from a singularity that is still expanding (see Redshift)

Cadaver-a dead body used for research and study

Captain's Companion-an unpaid position in the early British navy with the purpose of keeping the captain company on long voyages

Carbon 14-a radioactive version of carbon that has a hypothetical half-life of about 5,730 years

Catastrophism-large changes to a landscape over a very short period of time (see Gradualism)

Cholera-a serious disease that can be fatal and is spread by untreated water

Chromosome-a complex concentration of that are organized together to facilitate cell division (see gene)

Class-the phylogenetic classification stage between phylum and family

Coccyx-a bone just below the pelvis that helps with balance

Coelocanth-a large fish that lives off the coast of Madagascar

Colony-a collection of organisms of the same species that come together for a mutual purpose, often with specialized roles

Condensation-amino acids or other macromolecules coming together to form more complex structures and releasing water (see Hydrolysis)

Daughter Compound-the resulting element after undergoing radioactive decay (see Parent Compound)

Dinosaur-collective term for large reptilian creatures thought to be extinct

Element-a class of substances that are only made of one type of atom

Empirical-data which is based in reasoning from observation

Epiglottis-a flap that is present above the wind pipe to keep food and water from going into the lungs

Erector Pili-structures in the skin that help to raise hair (commonly referred to as goosebumps) under certain conditions

Eugenics-the policy of eliminating the people that are considered to have less desirable traits

Evolution-the hypothesis that all creatures on Earth developed from a single common ancestor

Experimental Science-the area of science that deals with observable and testable research (see Historical Science)

Extant-a species of organisms that is still living

Extinct-a species of organisms that is no longer living

Family-the phylogenetic classification state between class and genus

Forensics-the branch of historical science that deals with the reconstruction of events at a crime scene

Fossil-a dead organism that was quickly covered with sediments that fused into the organism
Galapagos Islands-a group of islands off the coast of South America, often connected with the birth of the evolutionary hypothesis
Gene-a single trait that is inherited from an organism's parents
Geocentric-the false hypothesis that the Earth is the center of the universe (see Heliocentric)
Geologic Clock-methods to measure how old the Earth is by studying the rate of natural phenomenon
Geologic Column-a hypothetical ordering of fossils by strata that shows the different organisms of Earth's history
Geology-the branch of science that deals with the rocks and the Earth's surface
Germ Theory-a medical theory that states that microscopic organisms are the primary cause of disease (see Miasma Theory)
Gill Slits-hypothetical organs that humans are said to have in the womb that resemble the gills of fish
Gradualism-slow incremental changes to a landscape over a long period of time (see Catastrophism)
Half-Life-the time that it takes for half of a radioactive compound to decay (see Radioactive Dating)
Heliocentric-the theory that the Sun is the center of the solar system (see Geocentric)
Historical Science-a term used to refer to the area of science that specifically deals with the recreation of events (see Experimental Science)
Homology-the belief that common bone structures indicate a common ancestor
Horse Chesnut-a supposed vestigial organ in horses that is present on the inside of the front legs
Hydrolysis-proteins or other macromolecules breaking apart to form simple compounds and taking in water (see Condensation)
Hypothesis-an early step in the scientific method in which the researcher states what he expects to see before beginning an experiment
Index Fossil-a fossil that is associated with a particular age range in the geologic column and is used to date the rocks
Inorganic-any collection of molecules that does not exhibit the signs of life and was never alive
Junk DNA-components of DNA that hypothetically serve no purpose (see telomere)
Kongamato-a large creature that lives in Africa that terrorizes the local populace and shares many physical traits with pterodactyls
Law-a theory that has been proven true so often that it is undeniable and is generally associated with mathematical proofs (see Theory)
Miasma Theory-a medical hypothesis that falsely argued that bad air was the primary cause of disease (see Germ Theory)
Microbiology-the branch of science that deals with microscopic or single-celled organisms
Microscope-a device for viewing organisms that are too small to be seen with the naked eye
Millions of Years-a principle vital to evolution that posits that the origin of the universe, Earth and life itself takes place over the course of a very long period of time
Naturalism-the branch of science that deals with the natural world and the systems that it contains
Nebula-a cloud of gas and dust that exists in space
Non-Living-a collection of molecules that once exhibited the signs of life, but no longer does

Nuclear Fusion-the process by which atoms fuse together under extreme heat and pressure to make heavier elements
Nucleotide-the unit on the DNA that acts as a single letter on the genetic code
Oort Cloud-A hypothetical area of space outside of our solar system that generates comets
Organic-any collection of matter that presently exhibits all of the signs of life
Pakicetus-a hypothetical ancestor of aquatic mammals
Paleontology-the branch of science that deals with fossils and prehistorical research
Pangea-a hypothetical landmass in which all the continents were one and gradually broke apart
Parent Compound-the original element before undergoing radioactive decay (see Daughter Compound)
Parallax Trigonometry-trigonometry that measures the locations of astral bodies up to 100 light years away
Phylogenetic-the classification system of animals
Polonium-a radioactive element that decays very quickly
Polystrata Fossil-a fossil that extends through multiple layers of strata
Potassium-Argon-a radioactive dating method that tracks the decay of potassium into argon
Protein-a systematic collection of amino acids that act as one of the primary building block of life
Proto-Feather- a hypothetical tissue that is a transition between scales and feathers
Proxima Centauri-hypothetically one of the nearest stars to Earth
Radioactive Dating-the hypothetical process by which scientists can measure the age of a substance by looking at the decay rate (see Half-Life)
Redshift-a hypothetical method of measuring ever expanding universe (see Big Bang Hypothesis)
Salinity-the concentration of salt in water
Shuffling-the changes in already present traits in an organism due to environmental stress (see variety)
Sickle Cell Anemia-a mutation to red blood cells that causes them to take on a crescent shape
Simian-any ape-like creature
Single Common Ancestor-a hypothetical microbe from which all life descends
Soft Tissue-any non-skeletal tissue
Solar Eclipse-event in which the moon is directly in front of the sun from the vantage point on Earth
Sponge-a sedentary sea animal that is little more than specialized tissue
Stellar Nursery-hypothetical celestial structure that forges stars
Strata-layers of dirt and sediment that hypothetically show the different ages of rocks and fossils (see Geologic Column)
Super Nova Remnant-the remains of a large star after it explodes (dies)
Super Nova-a large stellar explosion caused by a star dying
Tectonic Plate-structures of the Earth's crust that are split up along fault lines
Telomere-a component of the DNA near the end of the strand that protects the cell from degradation from repeated copying (see Junk DNA)
Theodolite-a device used on Earth to measure the distance between objects via trigonometry
Theory-a hypothesis that has been proven true, but may still be disproven over time (see Law)
Transitional Form-a hypothetical organism that shares traits with both the organism that it descends from and the organism which it ascends to

Trigonometry-the branch of mathematics that deals with the measurement of triangles
Universal Common Ancestor-a hypothetical single-celled organism that was the first living organism on Earth
Variety-used to describe species and sub-species that differ only by minor differences (see shuffling)
Vestigial Organ-a structure that hypothetically serves no purpose in humans or any other creature
Vestigial Tail-a structure that humans have in the womb that is not actually a tail
Webbed Digits-tissue that remains between the fingers and toes after birth
Wisdom Tooth-a tooth at the back of the jaw that sometimes must be removed due to infection or a lack of space
World View-the presuppositions and biases that exist in all people and are ever present in discussions and debates